Fortschritte der Chemie organischer Naturstoffe

Progress in the Chemistry of Organic Natural Products

57

Founded by L. Zechmeister
Edited by W. Herz, G. W. Kirby, W. Steglich,
and Ch. Tamm

Authors:
E. Casadevall, D. P. Chakraborty, C. Largeau,
P. Metzger, G. R. Pettit, S. Roy

Springer-Verlag
Wien New York 1991

Prof. W. HERZ, Department of Chemistry,
The Florida State University, Tallahassee, Florida, U.S.A.

Prof. G. W. KIRBY, Chemistry Department,
The University, Glasgow, Scotland

Prof. Dr. W. STEGLICH, Institut für Organische Chemie und Biochemie der Universität
Bonn, Bonn, Federal Republic of Germany

Prof. Dr. CH. TAMM, Institut für Organische Chemie der Universität Basel,
Basel, Switzerland

© 1991 by Springer-Verlag/Wien
Softcover reprint of the hardcover 1st edition 1991
Library of Congress Catalog Card Number AC 39-1015

Typesetting: Macmillan India Ltd., Bangalore-25

Printed on acid free paper

With 2 coloured Plates and 26 Figures

ISSN 0071-7886
ISBN-13:978-3-7091-9121-7   e-ISBN-13:978-3-7091-9119-4
DOI: 10.1007/978-3-7091-9119-4

Contents

List of Contributors

CASADEVALL, Dr. E., Laboratoire de Chimie Bioorganique et Organique Physique, École Nationale Supérieure de Chimie de Paris, 11, Rue Pierre & Marie Curie, F-75231 Paris Cedex 05, France.

CHAKRABORTY, Prof. D. P., 11/1/5, Satchasipara Lane, Calcutta 700036, India.

LARGEAU, Dr. C., Laboratoire de Chimie Bioorganique et Organique Physique, École Nationale Supérieure de Chimie de Paris, 11, Rue Pierre & Marie Curie, F-75231 Paris Cedex 05, France.

METZGER, Dr. P., Laboratoire de Chimie Bioorganique et Organique Physique, École Nationale Supérieure de Chimie de Paris, 11, Rue Pierre & Marie Curie, F-75231 Paris Cedex 05, France.

PETTIT, Prof. G. R., Cancer Research Institute, Arizona State University, Tempe, AR 85287-1604, U.S.A.

ROY, Dr. S., 10, J. N. Roy Lane, Calcutta 700006, India.

Lipids and Macromolecular Lipids of the Hydrocarbon-rich Microalga *Botryococcus braunii*. Chemical Structure and Biosynthesis. Geochemical and Biotechnological Importance

P. Metzger, C. Largeau, and E. Casadevall

Laboratoire de Chimie Bioorganique et Organique Physique, Ecole Nationale Supérieure de Chimie de Paris, CNRS, France

With 2 Plates and 19 Figures

Contents

I. Introduction

Botryococcus braunii Kützing is a colonial Chlorophyceae (green microalga) characterized by an unusually high production of lipids and an original organization of colonies. This species is widely distributed on all continents, in freshwater, brackish and saline lakes, reservoirs or even small pools, situated in temperate, tropical and continental zones as well (*1*). The ability to develop spectacular blooms on the surface of unruffled waters which consist of a floating mass of colonies rich in oil is also a conspicuous feature of *B. braunii*. Thus during a survey of the 4000 ha Darwin River Reservoir in Australia, Wake and Hillen (*2*) estimated a bloom of *B. braunii* at 1500 tons and assaying at 30% oil.

On observation under a light microscope, *B. braunii* colonies appear as clusters of cells which are variable in size, sometimes up to 1 mm; several clusters can be attached to one another by transparent and refringent threads (*3*). The pyriform shaped cells (usual size 13 μm × 7 μm) are embedded in a matrix impregnated by a pale yellow to red coloured oil which, by pressure on the cover-slip, can be excreted from the colonies. In aged cells, chloroplasts exhibit starch grains, a typical feature of Chlorophyceae, and the cytoplasm is invaded by lipid inclusions. Transmission electron microscopy (Plate 1) revealed the structure of the thick matrix surrounding the basal part of cells (*4, 5*). This matrix consists of outer walls originating from successive cellular divisions; closely adhering to each other these outer walls ensure colony cohesion.

While the very high oil production typical of *B. braunii* has been recognized for almost a century (*6*), the nature of the lipids forming this oil which are easily extracted by organic solvents remained obscure for a long time. According to the work of Zalessky in 1926 (*7*) and Blackburn in 1936 (*3*), *B. braunii* oil was at first assumed to be composed of fatty acids. This concept was changed only in the 1960's when Swain and

GILBY (*8*) and MAXWELL and co-workers (*9*) finally established that the oil mainly contained hydrocarbons. Afterwards the study of *B. braunii* was somewhat complicated by BROWN et al. (*10*) who discovered the existence of green active and orange resting colonies, each characterized by the production of hydrocarbons of a very different nature, thus presenting an intriguing problem of metabolic switchover. Indeed, BROWN postulated that the green active form exclusively synthesized straight chain hydrocarbons, odd carbon numbered C_{23}–C_{31} alkadienes and trienes, whereas the orange resting colonies produced only a class of C_{30}–C_{37} triterpenes termed botryococcenes. However, an interconversion between these two physiological forms was never observed either directly in nature or in laboratory cultures, whatever the growth conditions were. The situation was clarified through the isolation of a number of strains of different geographical origin (Table 1). Each strain was thus shown to produce

Table 1. *Hydrocarbon Types and Geographical Origin of B. braunii Samples*

Hydrocarbon types	Sources and References
n-Alkadienes and trienes, odd from C_{23} to C_{31}; *n*-alkenes: minor, rarely found	– Freshwaters – Australia: Tarago, Sorrento (*13*); Bolivia: Challapata, Overjuyo, Pata Khota (*14*); England: Maddingley Brick Pits* (*15*)
(A race)	France: Chaumeçon, Coat ar Herno, Crescent, Grosbois, Lingoult (*11, 14, 16*) Morocco: Oukaimden (*11*)
	– Brackishwater – Bolivia: Titicaca (*14*)
Botryococcenes C_{30}–C_{37} triterpenes	– Freshwaters – Australia: Darwin Reservoir, Devilbend, Green Lake, McCay's Reservoir (*13*);
(B race)	Bolivia: Overjuyo (*17*); England: Oak Mere (*9*); France: Pareloup, Sanguinet, Vioreau (*11, 17*); Ivory Coast: Ayame, Taabo, Yamoussoukro (*17*); Philippines: Katugday (*17*); Thailand: Khao Kho Hong (*17*); USA: Lake Michigan (*18*), Berkeley (*12*); West Indies: La Manzo, Paquemar (Martinique), Chateaubrun (Guadeloupe) (*11*).
Lycopadiene C_{40} tetraterpene (L race)	– Freshwaters – Ivory Coast: Kossou, Yamoussoukro (*19*); Thailand: Khao Kho Hong (*19*).

* The strains available from the following culture centers: Cambridge (UK), Göttingen (FRG), Austin (Texas, USA) and Thonon (France), are all derived from the same sample collected in this lake.

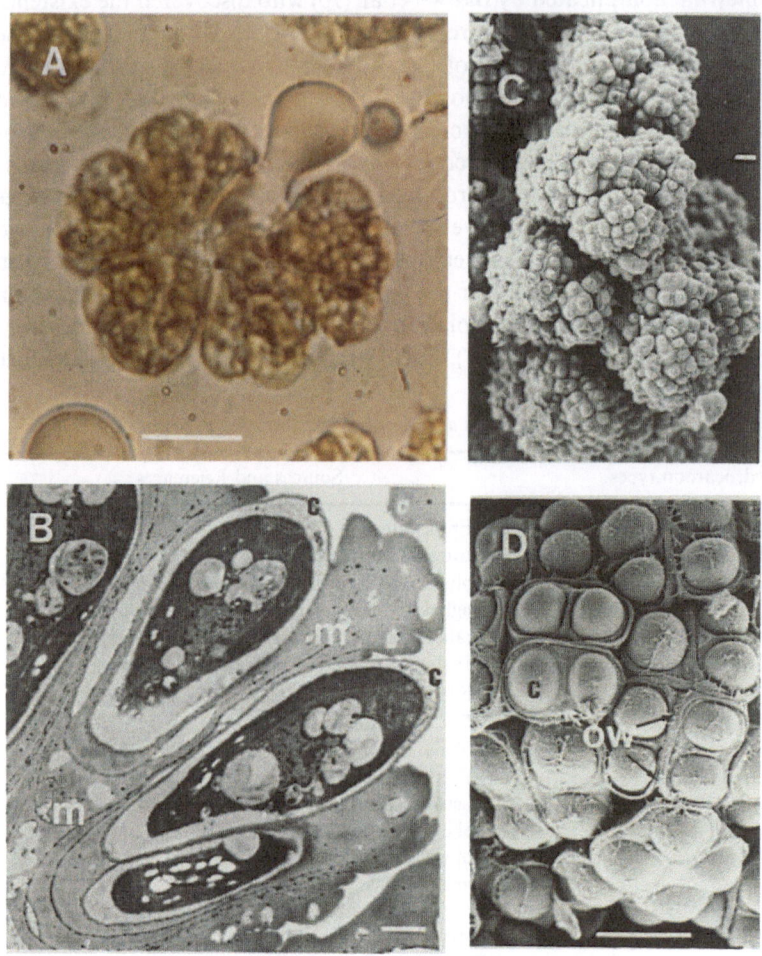

Plate 1. Micrographs of extant *B. braunii*. A: light microscopy of the Cambridge strain (A race); a refringent glouble of lipids is excreted from the colony by pressure on the coverglass. B: transmission electron microscopy of the Yamoussoukro strain (L race); the longitudinal section shows numerous lipid inclusions within the cells; the successive outer walls form a dense matrix (*m*). C and D: scanning electron microscopy of a Martinique strain (B race). C: whole colony. D: the basal part of the cells is embedded in thick outer walls (*ow*) when the apical part is covered by a thin cap (*c*). Scale bars: A, C and D 10 μm; B 1 μm. By courtesy of Dr. Berkaloff, ENS Paris (B), and Dr. Coute, Museum Paris (C and D)

References, pp. 63–70

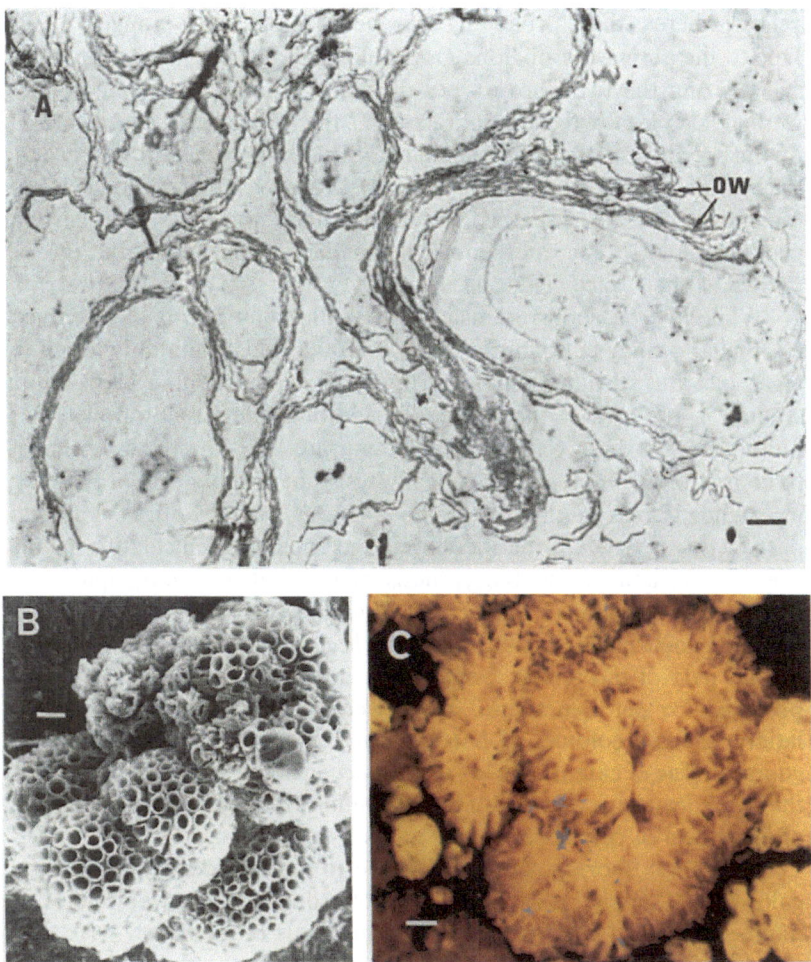

Plate 2. A: transmission electron microscopy of the final resistant material after chemical treatments of the Cambridge strain (A race); all the cell contents have disappeared, the empty areas are surrounded by successive PRB-composed outer walls (*ow*). B: scanning electron microscopy of a *B. braunii* colony in a recent (40,000 years) sediment core from East Africa; the cell contents have been entirely removed, on the contrary the outer walls are well preserved. C: UV fluorescence microscopy of fossil colonies of *B. braunii* from an immature Permian (*ca* $180 \cdot 10^6$ years) sedimentary rock from Australia; the general organization of the colonies is still well recognizable. Scale bars: A 1 µm, B and C 10 µm. By courtesy of Dr. BERKALOFF, ENS Paris (A), Dr. COUTE, Museum Paris (B) and Dr. PENIGUEL, SNEA Pau (C)

only one type of hydrocarbons at any growth stage; in old cultures the algae synthesizing alkadienes and trienes turned pale green or pale yellow, whereas those producing botryococcenes became orange (*11, 12*). Due to the absence of obvious morphological differences between algae synthesizing the two above types of hydrocarbons, *B. braunii* isolates were divided into two chemical races (*11*): the A race being characterized by the production of straight chain alkadienes and trienes, the B race by the synthesis of botryococcenes (Table 1). More recently, a third chemical race of *B. braunii* was recognized, in which the sole hydrocarbon produced is a C_{40} tetraterpenoid, lycopadiene, from which the race is known as the L race (Table 1) (*19, 20*).

While hydrocarbons are usually the main components of *B. braunii* oil, the latter may contain not only the usual lipids commonly found in algal oils but also non-classical lipids which have been thoroughly studied in the case of the A race. In addition, the structural elements building up the outer walls of the different races of *B. braunii* were shown to consist of insoluble and chemically resistant biopolymers originating in extended cross linking of the lipids through ether bridges. The first part of this review will be therefore concerned with elucidation of the chemical structures and biosynthesis of the hydrocarbons, lipids and macromolecular lipids produced by *B. braunii*.

The geochemical significance of *B. braunii* will be discussed in the second part of the review. Some sedimentary rocks, with a very high potential for petroleum production were shown to contain accumulations of fossil colonies of *Botryococcus*. Comparison of the chemical compositions of extant and fossil *Botryococcus* allows us to account for the above feature and to derive important information on the mechanism of formation of petroleum source rocks.

The energy shortage of the 1970's led to the consideration that mass culture of *B. braunii* might be a possible source of renewable fuels. A number of studies summarized in the third part of the review were thus concerned with the influence of culture parameters on the hydrocarbon production of *B. braunii* and with the optimization of this production.

II. Structure, Abundance and Biosynthesis of Lipids and Macromolecular Lipids in *B. braunii*

The bulk of *B. braunii* lipids is located in the outer walls which surround the basal part of the cells and build up the matrix of the colonies. Studies on the A and B races, including *in vivo* examination by

Raman microprobes revealed that *ca* 95% of the hydrocarbons are stored in the outer walls (*12, 21–23*). These external lipids are easily recovered by short extraction of the dry biomass with hexane as a yellow to red, sometimes brownish, coloured oil. In contrast, recovery of the internal lipids which are located in the membranes and in cytoplasmic inclusions requires prolonged extraction with $CHCl_3$–MeOH (*12, 21*).

Lipid analysis of the three races of *B. braunii* indicated, along with large amounts of specific hydrocarbons, the presence of common neutral lipids, such as triacylglycerols, sterols and carotenoids whose distribution will be considered shortly. In addition, extensive analytical investigations on the external lipids of various strains of the A race resulted in the identification of several series of non-classical compounds such as aldehydes, alkenylphenols, epoxides and ether lipids.

The nature of the insoluble biopolymers building up the outer walls in the three races was also studied. Contrary to the constituents of inner walls, these biomacromolecules are not polysaccharidic and show a very high resistance to drastic, non-oxidative chemical treatments. This type of biopolymer termed PRB (for "polymère résistant de *Botryococcus*") is isolated from the lipid-free biomass by means of drastic basic and acid hydrolyses.

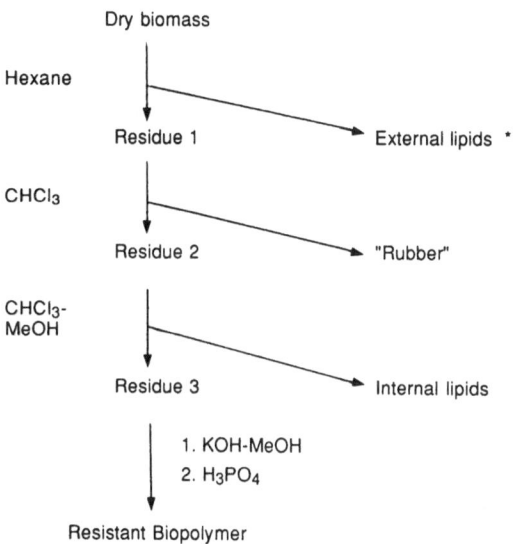

* The external lipids comprise the main part of hydrocarbons and in the case of the A race compounds exhibiting classical and non-classical structures.

Fig. 1. Isolation procedure of lipids and macromolecular lipids from *B. braunii*

It was recently observed that an additional extraction by $CHCl_3$, carried out between the hexane and $CHCl_3$–MeOH treatments provided a rubbery material from the three races of *B. braunii*. In the absence of this further extraction, "rubber" remained included, along with PRB, in the final, hydrolysis resistant, fraction. Structural and biosynthetic considerations to be developed in this review lead us to regard the PRB and *B. braunii* "rubbers" as macromolecular lipids.

The procedure used for the isolation of the lipids, "rubber" and PRB from *B. braunii* is described in Fig. 1. Mass spectrometry, IR, ^{1}H- and ^{13}C-NMR spectroscopy were used extensively for structure determination of the lipids and macromolecular lipids isolated from *B. braunii*. In this review, only the main spectral data for individual compound identification will be discussed. Hydrogenation, pyrolysis and chemical degradation by ozonolysis also contributed to structure elucidation. Investigations of the biosynthetic pathways were performed by feeding the algae with ^{2}H, ^{3}H, ^{13}C and ^{14}C labelled precursors. Radioactive labelling was examined by means of radio-GC and scintillation counting of HPLC fractions; ^{2}H and ^{13}C NMR spectrometry permitted determination of the labelling patterns obtained with ^{2}H and ^{13}C labelled precursors.

1. Hydrocarbons

Hydrocarbons generally constitute the main components of *B. braunii* external lipids and for a given strain the same type of hydrocarbons occurs in the external and internal lipid pools. These compounds are easily separated by SiO_2 or Al_2O_3 column chromatography, using hexane as eluent. When required, the resolution of hydrocarbon fractions into individual components can be achieved by HPLC.

1.1 Straight Chain Hydrocarbons (A Race)

1.1.1 Isolation and Structure Determination

The first structural study was concerned with the hydrocarbons of a strain from the culture center at Cambridge. C_{27}, C_{29} and C_{31} dienic compounds, series (1), were thus shown (15) to account for more than 90 percent of the hydrocarbons extracted from *B. braunii* biomass with acetone (it was observed in subsequent studies (21) that only external hydrocarbons are recovered using this solvent). Ozonolysis of the hydrocarbon mixture and analysis of the resulting aldehydes by GCMS indicated that the three above dienes exhibit a terminal and $C(\omega9)-C(\omega10)$ double bonds. Moreover *cis* stereochemistry for the internal unsaturation was suggested by an IR band at 720 cm^{-1}; this stereochemistry was further confirmed by ^{13}C NMR analysis (*16*).

In addition to this *cis* series, a second dienic series (2) was first identified in a French strain originating from Lake Grosbois (*16*). Freezing an hexane solution of the external hydrocarbons of this strain led to the separation of a solid mixture comprising the major compounds of the second dienic series. Oxidative and spectroscopic investigations indicated the same general structure as in series (1), but with *trans* stereochemistry (IR stretch at 970 cm^{-1}) of the $C(\omega9)-C(\omega10)$ double bond in the compounds of series (2).

$$CH_3 - (CH_2)_7 \underset{\omega9}{\overset{E \text{ or } Z}{- CH}} = \underset{\omega10}{CH} - (CH_2)_x - CH = CH_2 \qquad \begin{matrix}(1) \ (Z) \\ (2) \ (E)\end{matrix} \qquad x = odd, 11\text{–}19$$

$$CH_3 - (CH_2)_5 \underset{\omega7}{\overset{Z}{- CH}} = CH - \underset{\omega9}{\overset{Z}{CH}} = CH - (CH_2)_x - CH = CH_2 \qquad \begin{matrix}(3) \ x=17 \\ (4) \ x=15\end{matrix}$$

Reversed-phase HPLC of the external hydrocarbon fraction from a second collection strain obtained from the culture center at Austin furnished a mixture of a C_{29} triene and a C_{27} diene (C_{18} column; acetone/acetonitrile/THF 5:12:1 v/v/v) (*16*). Further separation of this eluate on a silver nitrate-silica gel column afforded a subfraction enriched in triene. Its IR spectrum exhibited two bands at 910 and 990 cm^{-1} indicative of a non-conjugated terminal double bond. The existence of two conjugated double bonds was evident from an UV absorbance at 232 nm. Ozonolysis, followed by oxidative cleavage of the resulting ozonide gave nonadeca-1,19 dioic acid generated by $C(1)-C(2)$ and $C(20)-C(21)$ cleavages. This result established that triene (3) is a nonacosa-1,20,22 triene. The low coupling constant observed in the ^1H-NMR spectrum between the olefinic protons (7.5 Hz) revealed the *cis, cis* stereochemistry of the conjugated double bonds. Recently, the structure of a lower homologue, the C_{27} triene (4) which is an important constituent of the external hydrocarbon fraction isolated from another French strain (Lake Coat ar Herno) (*14*) was determined; the conjugated double bonds at $C(\omega7)$ and $C(\omega9)$ were shown to exhibit also *cis, cis* stereochemistry (*24*).

The mono-unsaturated odd carbon numbered $C_{23}-C_{27}$ hydrocarbons which occur as minor constituents in a very limited number of strains, for example in the Coat ar Herno strain, were examined by GCMS. Terminal position of the unsaturation was evident from the mass spectra which contained a $[M-28]^+$ ion (25).

The synthesis of only one alkadiene of B. braunii has been performed: electrochemical coupling of undecen-11-oic acid with oleic acid via a Kolbe reaction afforded the heptacosa-1,18(Z) diene (1) (26).

In relation of their biosynthesis, it is interesting to note that all the hydrocarbons from the A race possess an odd carbon number and a terminal double bond; moreover all the dienes and the two trienes identified up to date exhibit $C(\omega 9)–C(\omega 10)$ unsaturation.

1.1.2 Biosynthesis

Because of their non-isoprenoid structure, the B. braunii dienes and trienes were expected to derive from fatty acids. This assumption was confirmed by feeding with $[16-^{14}C]$ palmitic, $[18-^{14}C]$ stearic and $[10-^{14}C]$ oleic acids, followed by analysis of the labelled hydrocarbons (27). In addition, the substantially higher incorporation level observed with oleic acid compared with incorporation in the saturated acids as well as the existence of a cis $C(\omega 9)–C(\omega 10)$ double bond both in the major hydrocarbons produced by the tested strains and in oleic acid suggested that the latter is the direct precursor of these hydrocarbons. The inverse relationship existing between oleic acid and the hydrocarbon abundances supported this assumption; indeed the oleic acid content of the algae remained low as long as hydrocarbon production was high, but rose considerably when hydrocarbon production sharply decreased.

Two distinct mechanisms starting from fatty acids are known to be implicated in straight-chain hydrocarbon biosynthesis:

(i) The elongation–decarboxylation mechanism in which a direct precursor, generally a C_{16} or C_{18} fatty acid derivative, is elongated by successive addition of C_2 units derived from malonyl CoA. When the appropriate chain lengths are reached, the very long chain fatty acid derivatives are decarboxylated to yield the corresponding hydrocarbons;

(ii) the head-to-head condensation of two fatty acid derivatives. In this mechanism, one of the acid derivatives is decarboxylated specifically following the condensation step, while the total carbon chain of the other is incorporated into hydrocarbons.

Feeding experiments with doubly labelled $[9,10-^3H, 1-^{14}C]$ and $[9,10-^3H, 10-^{14}C]$ oleic acid showed the complete incorporation of the chain of this acid into hydrocarbons. This, and the extensive inhibition of hydrocarbon synthesis in the presence of dithioerythritol (an inhibitor of decarboxylation in higher plants where hydrocarbon biosynthesis takes place *via* an elongation–decarboxylation pathway) established that hydrocarbon formation in *B. braunii* occurred following this latter mechanism (*27*).

However, a comparative study of the inhibition of hydrocarbon and very long chain fatty acid (VLCFA) synthesis by trichloroacetic acid, demonstrated that two distinct systems are involved in the formation of very long chain lipids in the A race of *B. braunii* (*28*). The bulk of VLCFA originates from an elongation pathway independent of the one leading to hydrocarbons. Only a minor part of total VLCFA is formed by release from the elongation–decarboxylation complex (Fig. 2); such distinct elongation pathways were also shown to occur in higher plants.

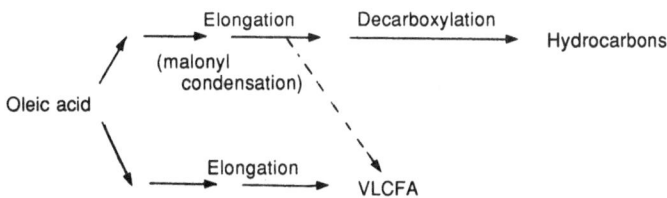

Fig. 2. Different pathways in chain elongation in the A race

Regarding the decarboxylation step, it is well documented that CO_2 elimination from carboxylic acids requires high energy and has to be activated by a β substituent which can stabilize the negative charge generated by CO_2 release. Accordingly it has generally been thought that activated fatty acyl derivatives are the intermediates in the decarboxylation leading to hydrocarbons. Confirmation for this assumption was obtained by feeding *B. braunii* with very long chain β-keto and β-hydroxy fatty acids (**5**) and (**6**), synthesized as shown in Fig. 3 (*29*). A more efficient label incorporation into dienic hydrocarbons (*ca* × 2) was observed from the C_{28} β-hydroxy ester than from its β-keto counterpart. These incorporations definitely ruled out a head-to-head condensation process and were consistent with an elongation–decarboxylation mechanism. The preferential labelling observed for the C_{27} diene, compared with the C_{29} and C_{31} dienes could originate from the direct decarboxylation of the bulk of the C_{28} precursor before subsequent elongation.

Considering these results and the mechanism of elongation of very long chain fatty acids, the activation step would not require the introduction of a specific substituent, the β-hydroxy group being normally formed during the elongation process. The systematic occurrence of a terminal

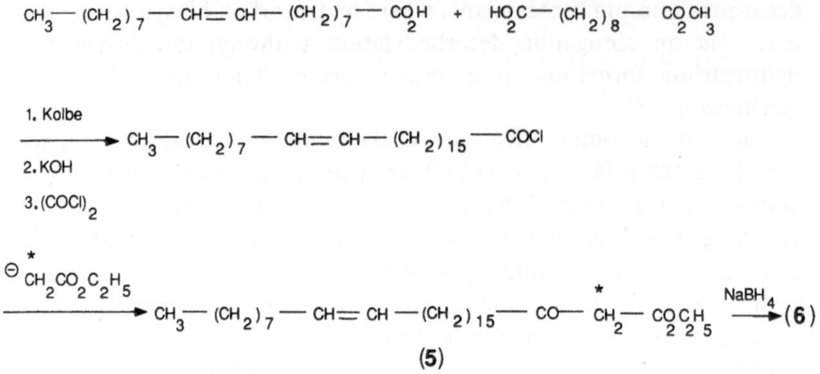

$$CH_3 - (CH_2)_7 - CH = CH - (CH_2)_7 - \underset{2}{C}\overset{O}{H} \;+\; H\underset{2}{O}C - (CH_2)_8 - \underset{2}{C}\underset{3}{O}CH$$

$$\xrightarrow[\substack{\text{2. KOH} \\ \text{3. (COCl)}_2}]{\text{1. Kolbe}} \; CH_3 - (CH_2)_7 - CH = CH - (CH_2)_{15} - COCl$$

$$\xrightarrow{\overset{*}{\underset{}{\overset{\ominus}{C}H_2 CO_2 C_2 H_5}}} \; CH_3 - (CH_2)_7 - CH = CH - (CH_2)_{15} - CO - \overset{*}{C}H_2 - \underset{2\,2\,5}{COCH} \xrightarrow{NaBH_4} (6)$$

$$(5)$$

Fig. 3. Synthesis of ^{14}C labelled β-keto and β-hydroxy C_{28} fatty acid derivatives (5) and (6)

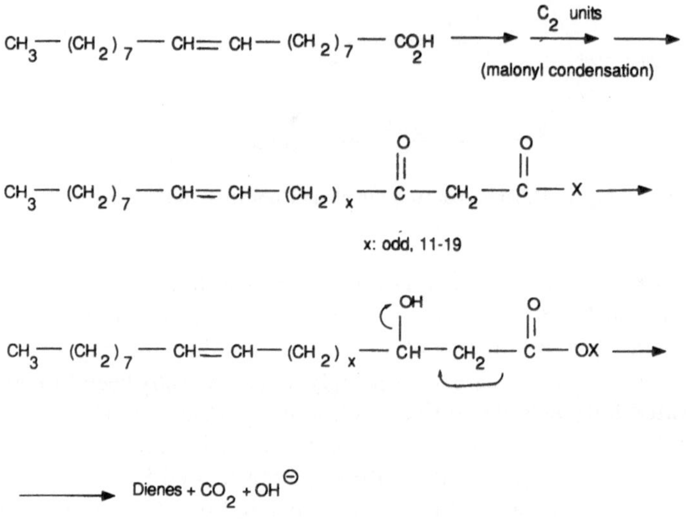

$$CH_3 - (CH_2)_7 - CH = CH - (CH_2)_7 - \underset{2}{C}\overset{O}{H} \xrightarrow[\text{(malonyl condensation)}]{C_2 \text{ units}}$$

$$CH_3 - (CH_2)_7 - CH = CH - (CH_2)_x - \overset{O}{\overset{\|}{C}} - CH_2 - \overset{O}{\overset{\|}{C}} - X \longrightarrow$$

x: odd, 11–19

$$CH_3 - (CH_2)_7 - CH = CH - (CH_2)_x - \overset{OH}{\overset{|}{C}H} - CH_2 - \overset{O}{\overset{\|}{C}} - OX \longrightarrow$$

$$\longrightarrow \text{Dienes} + CO_2 + OH^{\ominus}$$

Fig. 4. Biosynthesis of alkadienes from oleic acid

double bond at the "carboxylic" end of *B. braunii* hydrocarbons would be a direct consequence of CO_2 elimination *via* reductive decarboxylation of a β-hydroxy fatty acid derivative (Fig. 4).

Recently, the specificity of the elongation–decarboxylation system of *B. braunii* towards the formation of *cis* and *trans* dienes was tested by feeding experiments with radiolabelled oleic acid and with its *trans* isomer, elaidic acid (*30*). Although these experiments established that elaidic acid is the direct precursor of the *trans* hydrocarbons in *B. braunii*, this acid does not occur in appreciable amounts in algae synthesizing the *trans* dienes. In these strains, elaidic acid is formed by isomerization of oleic acid and subsequently converted at a relatively higher rate into *trans* dienes, so that it does not accumulate in detectable amounts. The absence of *trans* diene synthesis in other strains was shown to be related to the lack of a system capable to convert oleic acid into elaidic acid; indeed when elaidic acid was fed to these algae, *trans* dienes were synthesized. These results demonstrated the absence of specificity of the elongation–decarboxylation system regarding the configuration of the $C(\omega 9)$–$C(\omega 10)$ double bond in the C_{18} precursor acid.

1.1.3 Chemical Variability and Abundance of Hydrocarbons in the A Race

Twenty straight chain hydrocarbons have been so far identified on analysis of samples chiefly collected from freshwater lakes and of laboratory grown strains. As shown in Table 2 which describes the results corresponding to three cultivated strains, large variations can occur in both the nature and the relative abundance of these hydrocarbons.

WAKE and HILLEN (*13*) first analyzed several wild samples of the A race of *B. braunii*; they noted that hydrocarbon compositions in algae collected in two Australian lakes (Tarago and Sorrento) resembled that observed for a strain from the culture center at Cambridge (*15*), with essentially the odd carbon numbered C_{27}–C_{31} *cis* dienes (**1**) and the C_{29} triene (**3**). Such a hydrocarbon profile where C_{29} compounds predominate was also found in the collection strain from the culture center at Austin (Table 2) (*16*).

As previously discussed, the synthesis of *trans* dienes is related to the ability of some strains to isomerize oleic acid into elaidic acid, the direct precursor of *trans* isomers (**2**). In the case of the strain originating from Lake Titicaca, a brackish water lake, the *trans* dienes predominated strongly in the short chains and the *trans*/*cis* ratio progressively decreased with chain lengthening; on the whole a fifty-fifty repartition between the two dienic series was observed and no trienes were detected (Table 2) (*14*). The synthesis of *trans* dienes is not specific for *B. braunii* originating from brackish waters, indeed, as previously shown, the strain collected from the freshwater Lake Grosbois also synthesizes these hydrocarbons.

A large C_{27} predominance was observed in some strains, such as the one originating from the French Lake Coat ar Herno (Table 2); this strain is also characterized by a high proportion of trienes relatively to

Table 2. *Distribution (%) and Content of External Hydrocarbons in Three Cultivated Strains of the A Race*

Hydrocarbons	Location and stereochemistry of the double bonds	Origin		
		Collection strain Austin	Bolivia Titicaca	France Coat ar Herno
23:1	1	–	–	3.5
23:2	1, 14(Z)	trace	–	1.0
25:1	1	–	–	2.5
25:2	1, 16(Z)	0.4	0.1	2.5
25:2	1, 16(E)	–	0.7	–
25:3	n.d.	–	–	2.4
27:1	1	–	–	2.5
27:2	1, 18(Z)	9.9	3.1	31.0
27:2	1, 18(E)	–	13.9	–
27:3	n.d.	–	–	10.4
27:3	1, 18(Z), 20(Z)	–	–	25.2
27:3	n.d.	–	–	4.1
29:2	1, 20(Z)	43.8	29.3	11.4
29:2	1, 20(E)	–	32.6	–
29:3	1, 20(Z), 22(Z)	12.0	–	2.3
31:2	1, 22(Z)	28.8	19.3	–
31:2	1, 22(E)	–	3.8	–
31:3	n.d.	2.2	–	–
Others		2.9	1.2	1.2
Content (*)		20	12.8	1.6

n.d.: not determined.
Data taken from refs. *14* and *16*.
*Total external hydrocarbons as % of dry wt.

total hydrocarbons, substantial amounts of the rarely occurring mono-olefins and by the lack of C_{31} components (*14, 24*).

Due to the identical culture conditions used in the comparative study of Table 2 and to the similar physiological stage of the analyzed cells, it can be concluded that the observed chemical variations are under genetic control. In laboratory cultures hydrocarbon composition remained fairly constant except in triene-producing strains, where the level of trienic hydrocarbons slightly increased at the end of the active growth phase.

In the A race, the variations noticed for the hydocarbon content in relation with the strain origin, and for identical culture conditions, are very marked: levels ranging from *ca* 1% up to more than 60% of dry biomass were observed. The highest content was found in a French strain originating from Lake Chaumeçon, where external hydrocarbons reach the impressive level of 61% of dry wt (*11*); on the contrary in another French isolate, the Coat ar Herno strain (Table 2), they appear as minor components: 1.6% of dry wt (*14*). Generally, A strains exhibit intermediate hydrocarbon levels, for example 20% in the strain from Austin (Table 2). As discussed below (see 3.4), the weak hydrocarbon content of some strains could be related to the metabolism of these compounds into lipids of higher molecular weight such as some ether lipids.

1.2 Botryococcenes (B Race)

1.2.1 Isolation and Structure Determination

The first hydrocarbon described from the B race was a C_{34} compound called "botryococcene" (*31*), one of the most widespread members of the botryococcene family as demonstrated by subsequent studies. Due to its large predominance in the total hydrocarbon fraction isolated from some strains, *ca* 90%, no additional purification was needed for identification of the C_{34} compound. This simple case, however, occurs rarely and isolation of pure or at least enriched fractions of the other botryococcenes was required. This purification was obtained by reversed-phase HPLC using analytical C_{18} columns and by repeated injections of botryococcene mixtures; various eluents such as acetone-acetonitrile 2:3 (*32*), acetonitrile (*33*) and acetone-methanol 3:2 were used (*34*). This technique was however ineffective for separation of hydrocarbons exhibiting only regio- and (or) stereo-isomerism of double bonds.

Exposure of botryococcene-containing *Botryococcus* to the atmosphere for a few weeks is known to afford a brown elastic material which was termed "*Botryococcus* rubber" (*35*). From this material, botryococcenes were recovered in much smaller quantities when compared to extraction of fresh biomass; at the same time a complex mixture of oxidation products derived from these hydrocarbons was isolated and partial transformation of the botryococcenes into an insoluble abiotic polymer was observed (*36*). To avoid these transformations also noticed with pure compounds botryococcenes must be stored in hexane solution under a nitrogen atmosphere at $-25\,°C$.

To date, among the fifty botryococcenes detected by GC, sixteen structures have been determined. The main tools of the procedure will be described for botryococcene C_{30} which occupies a key position in the series. In addition, some detailed spectral data will be given for identification of higher botryococcenes in order of increasing structural complexity.

1.2.1.1. Botryococcene $C_{30}H_{50}$

Botryococcene C_{30} (7) was first identified in a strain originating from a West Indian lake situated in Martinique (Lake La Manzo; MLM3 strain in ref. *32*) where it accounted for 7% of the external hydrocarbon fraction. This compound isolated in pure form as a colourless liquid was optically active ($[\alpha]_D^{20} - 14°$, c 2.43 CHCl$_3$). The molecular composition was deduced as $C_{30}H_{50}$ from EIMS analysis; ^{1}H and ^{13}C NMR spectroscopy revealed the presence of six double bonds.

In addition, the ^{1}H- and ^{13}C-NMR spectra of (7) suggested a terpenoid nature. Thus the ^{1}H-NMR spectrum showed signals for eight methyl groups in the range δ 0.93–1.68, thirteen methylene and methine allylic protons at δ1.87–2.13 and nine olefinic protons at δ4.90–5.86; it also contained resonances for four nonallylic protons at δ1.25–1.43. The following partial structures could be identified by spin–spin decoupling and two dimensional ^{1}H–^{1}H correlations and confirmed by ^{13}C NMR analysis:

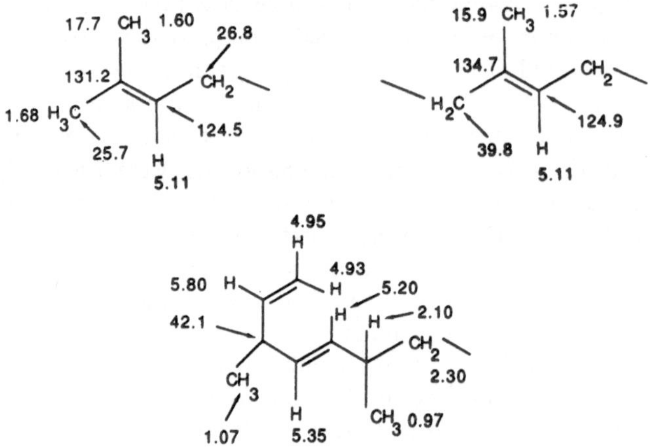

The existence of a quaternary center at C(10) was ascertained by EIMS fragmentation of the fully hydrogenated derivative, botryococcane $C_{30}H_{62}$ (8) (Fig. 5a). Ozonolysis of (7) followed by oxidative cleavage of the polyozonide gave rise to equal amounts of laevulinic acid (9) and 2-methyl glutaric acid (10) (Fig. 6), thus allowing to definitely establish the structure.

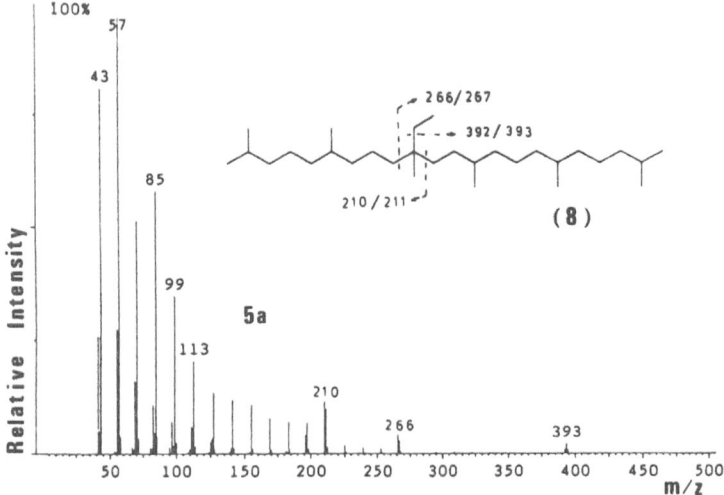

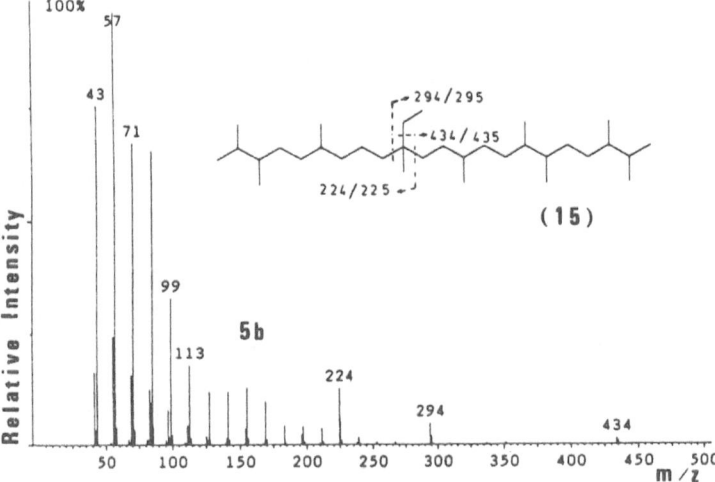

Fig. 5. EI-MS of botryococcanes $C_{30}H_{62}$ (8) and $C_{33}H_{68}$ (15)

The molecular composition of all the higher botryococcenes was shown by EIMS analyses to correspond to the general formula $C_{30+x}H_{50+2x}$, x = 1–7. The permanence of the central framework of (7) in all these structures was evident from the almost invariant C(9)–C(13) and C(25)–C(27) resonances in the ^{13}C-NMR spectra, while structural modifications occurred on the trisubstituted double bonds.

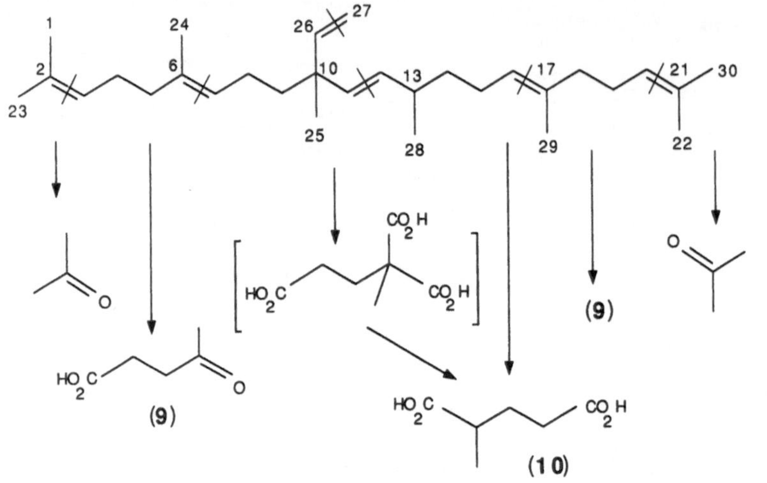

Fig. 6. Structure and ozonolysis of $C_{30}H_{50}$ botryococcene (7)

1.2.1.2 Higher Acyclic Botryococcenes

A C_{31} and a C_{32} botryococcene, respectively (11) and (12), were also isolated from the above Martinique strain (MLM3 strain) (32). On ^{1}H- and ^{13}C-NMR examination and by comparison with the C_{30} botryococcene (7), they exhibited one, respectively two, α,α'-disubstituted double bonds, with a proton signal at $\delta 4.60$ and ^{13}C resonances at $\delta 109.3$ and 150.0 instead of the trisubstituted unsaturation at C(2) or (and) C(20). Moreover, resonances for respectively one and two additional methyls (δ^{13}C 19.7) tied to methine carbons C(3) or (and) C(20) were also observed. The characteristic fragmentations of the derived botryococcane around the quaternary center at C(10) (ion doublets at m/z 280/281 and 210/211) indicated that the extra methyl group in (11) was introduced at C(20). Huang and Poulter recently described a C_{31} regioisomer (13) (37).

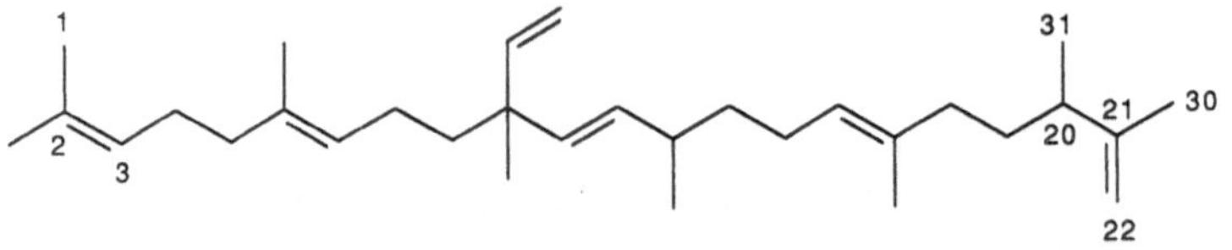

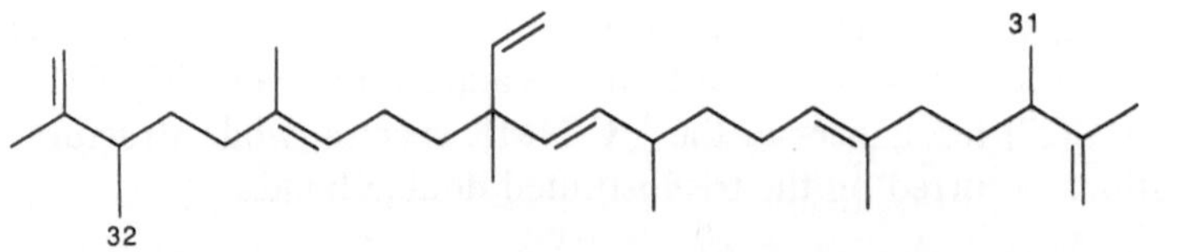

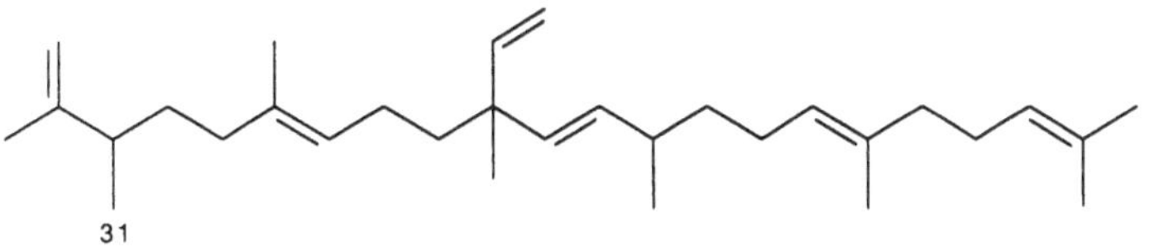

(13)

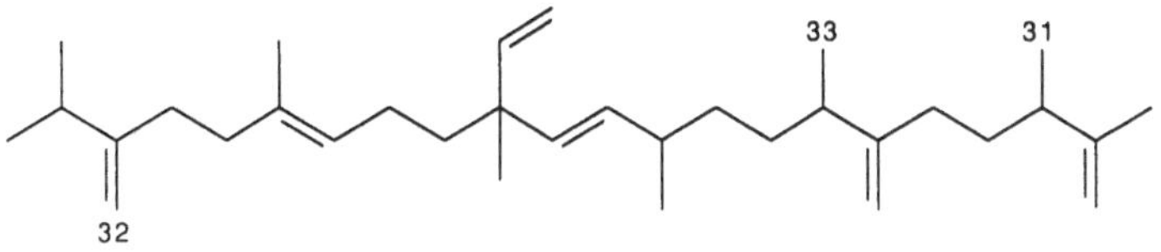

31

33   31

(14)

32

From a second Martinique strain also originating from Lake La Manzo (MLM1 strain) a C_{33} compound (14) was isolated as a major constituent (86%) of the external hydrocarbon fraction (*32*). In this case an additional methylation similar to those observed in (11)–(13) had occurred on the internal trisubstituted double bond at C(16). Furthermore, while alkylation at C(20) led to the formation of an isopropylidene group as observed in (11)–(13), alkylation at C(3) resulted in the formation of an isopropyl group in (14). The EIMS fragmentation of the derived botryococcane (15) (Fig. 5b) confirmed that methylations took place at C(3), C(16) and C(20) in (14).

The first discovered botryococcene, the C_{34} compound (16), was characterized by MAXWELL and co-workers in a wild sample collected in an English lake (Oakmere) (*9*). ^{13}C-NMR analysis and permanganate-periodate oxidation to (17) supported gross structure (16) (*31*). The absolute configuration of this botryococcene was recently established by oxidative degradation of the 26,27-dihydroderivative and comparison of the resulting enantiomers with optically pure synthons (*38*). The (S) configuration was thus ascribed to C(3, 7, 10, 16 and 20), while the asymmetric center C(13) was found to be (R).

A second C_{34} isomer (18) was isolated as a major constituent from a Martinique strain originating from a small pool in Paquemar (*32*); it accounted for 72% of the total hydrocarbon fraction. Structure (18) was proposed on the basis of NMR spectral evidence and ozonolysis degradation leading among other products to the ketoacid (17). The *cis* configuration of the trisubstituted C(5)–C(6) double bond was based on the ^{13}C chemical shift (δ34.5) of the allylic carbon C(7) which would be more deshielded in a *trans* configuration (δ39–40).

Structures of the C_{36} botryococcene (19) and of the C_{37} botryococcene (20) were established by ^{1}H- and ^{13}C-NMR studies and by EIMS of the derived botryococcanes (*32, 39*). These two compounds accounted for 35% respectively 10% of the external hydrocarbons extracted from an

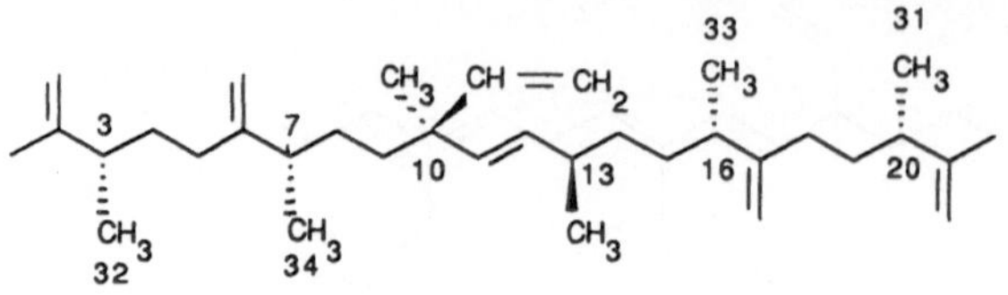

(16)

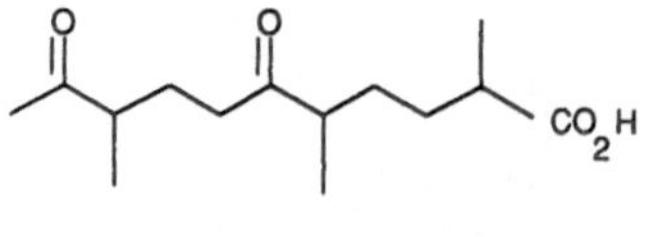

(17)

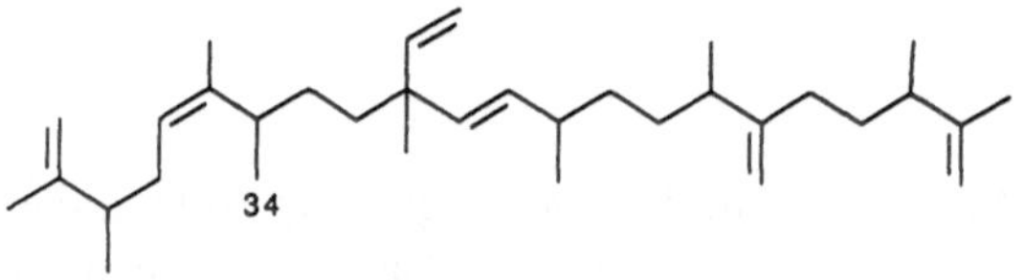

(18)

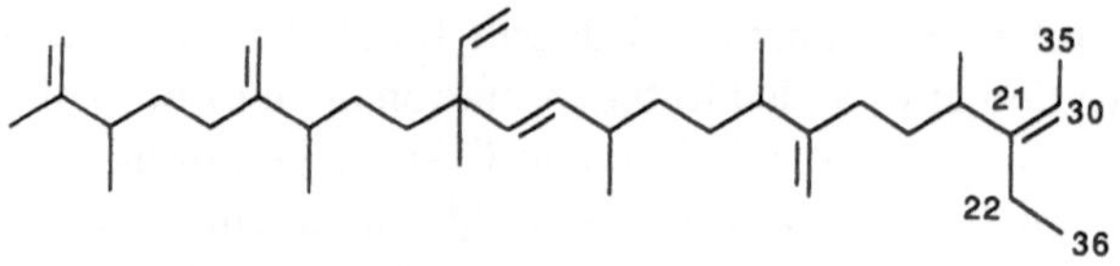

(19)

(20)

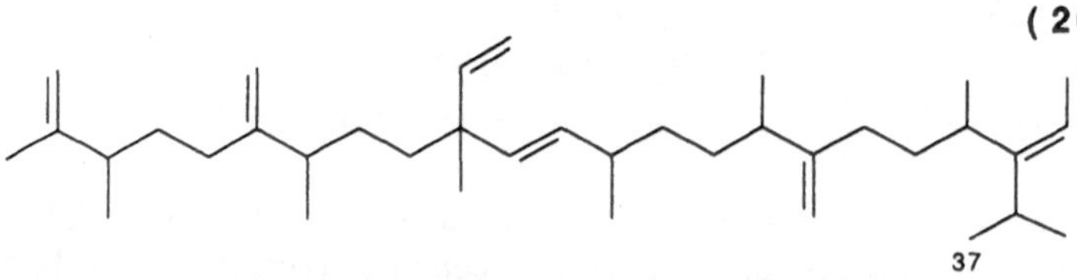

Australian sample collected in the Darwin Reservoir. The above compounds exhibited hypermethylation of one terminal isoprene unit, previously only exhibited by the methylated side chain of some sponge sterols (40).

1.2.1.3. Cyclobotryococcenes

The existence of a cyclic residue in some botryococcenes was evident from their ^{1}H- and ^{13}C-NMR spectra showing the presence of only five double bonds while their molecular ions indicated six degrees of unsaturation. Like their acyclic counterparts, all retained unchanged the central framework of the C_{30} botryococcene (7).

A C_{34} cyclobotryococcene (21) was isolated from the above Darwin sample as an important constituent (ca 15%) of the external hydrocarbon fraction (32). ^{1}H- and ^{13}C-NMR spectra and EIMS analysis of the derived cyclobotryococcane showed that structure (21) contains the C(1)–C(9) dimethylated moiety of the C_{34} compound (16) and a pentamethylated cyclohexenyl ring bound to C(15). Indeed, the ^{13}C-NMR spectrum showed the presence of a quaternary carbon (δ 34.5) bearing two geminal methyls (δ 19.8 and 29.5), one vinylic methyl (δ 24.8) and two methyls (δ 16.1 and 23.8) on methine carbons. The stereochemical features of this substituted ring have not yet been established.

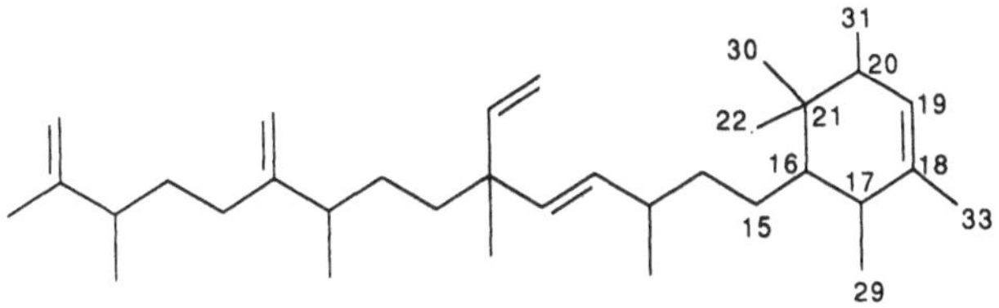

(21)

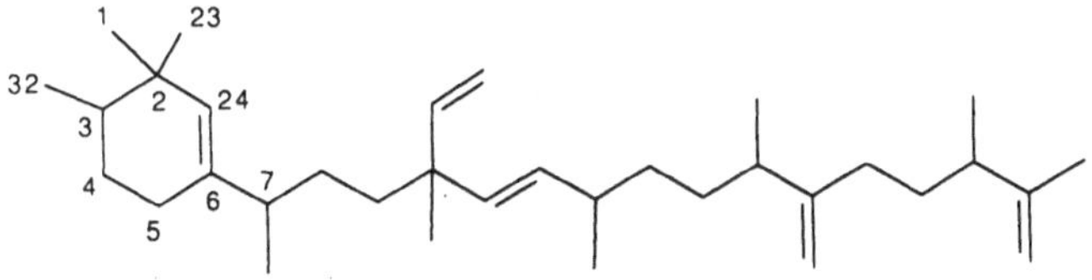

(22)

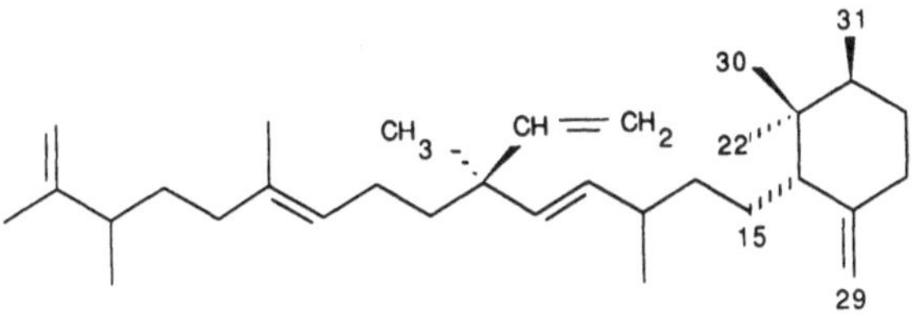

(23)

Another type of cyclohexane ring was found to occur in a second C_{34} cyclobotryococcene which accounted for half of the total hydrocarbons produced by a strain originating from Ivory Coast (Lake Ayame) (41). Chemical structure (22) was proposed on the basis of extensive ^{1}H- and ^{13}C-NMR analyses; however due to overlapping of some proton signals the geometry of the trimethylated cyclohexenyl ring could not be determined.

A C_{32} cyclobotryococcene (23) was described simultaneously by three groups working on two different strains. This hydrocarbon was isolated as an important constituent, 25% and 10% of the external hydrocarbons respectively, of the Berkeley strain (33, 34) and of a Bolivian strain originating from a lake situated in the Overjuyo valley (41). Intensive ^{1}H- and ^{13}C-NMR investigations as well as ^{1}H–^{1}H and ^{1}H–^{13}C correlations demonstrated the presence of a trimethylated methylenecyclohexane ring previously described in some *Iris* triterpenoids (42). NOE experiments gave evidence of an axial orientation of the large acyclic substituent, while Me(31) was equatorially oriented (*41*). This unexpected axial geometry removes the strong interactions existing for an equatorial orientation between H(15) and H(14) on one hand and H(29) and a methyl group on C(21) on the other. The stereochemistry at C(10) was shown to be (*S*) by degradation of (**23**) and comparison of the corresponding fragment with an optically pure synthon, and thus identical with that found for (**16**) (*33*). Moreover, the absolute stereochemistries of the two chiral centers in the methylenecyclohexane ring were also found to be (*S*) on the basis of ORD spectra and CD curves of the keto ester derived from oxidative cleavage of C(11)–C(12) and C(17)–C(29) double bonds (*43*). Three minor cyclobotryococcenes, two C_{31} compounds and a regioisomer C_{32} exhibiting the same trimethylated methylenecyclohexane ring were also isolated from the Berkeley strain (*44*).

1.2.2 Botryococcene Biosynthesis

On the basis of structural considerations it was at first proposed that the C_{34} botryococcene (**16**) arose from a tail-to-tail linkage of two dimethylated C_{15} units (*31*). However, characterization of numerous lower and higher homologues and feeding experiments with ^{14}C and ^{13}C labelled precursors suggest that the C_{30} compound (**7**), derived from the irregular coupling of two farnesyl pyrophosphates *via* presqualene pyrophosphate, is in fact the precursor of all the higher botryococcenes.

1.2.2.1 The Presqualene Pathway

The involvement of presqualene, the precursor of squalene, in botryococcene biosynthesis has been recently reviewed by POULTER in a particularly interesting paper about non-head-to-tail linkage in terpenes (*45*); therefore it will be only briefly discussed here. Indications of a biosynthetic pathway *via* presqualene pyrophosphate (**24**) were first given by the similar absolute configurations at C(10) in botryococcenes C_{34} (**16**) and C_{32} (**23**) and at C(3) in the cyclopropane moiety of presqualene (the carbon numbering of presqualene is given in Fig. 7). Evidence for such a pathway was obtained by incubation of *B. braunii* with (R)-[1-^2H] farnesol, (*S*)-[1-^2H] farnesol and (*S*)-[1-^2H, 1-^{13}C] farnesol, followed by ^2H and ^{13}C NMR investigations of the labelled botryococcenes (*46*). Previous studies revealed that synthesis of squalene (**25**) from presqualene (**24**) involves first the formation of cation (**26**), followed by 1′–2 and 1′–3 migrations and by hydride attack at C(1′). Regarding botryococcene formation, it was shown that the conformation of presqualene bound to botryococcene synthetase is similar to that in squalene synthetase and that cation (26) is also first formed. But in the botryococcene pathway the opening of the cyclopropyl ring occurs *via* an 1'–2 cleavage and hydride attack takes place at C(3') (Fig. 7). Accordingly, only a very small shift of the cofactor relatively to the C_{30} substrate in the botryococcene synthetase, near C(3'), and in the squalene synthetase, near C(1'), could explain the regioselective production of botryococcene C_{30} (7) or of squalene (25). Using this biosynthetic scheme, POULTER suggested that

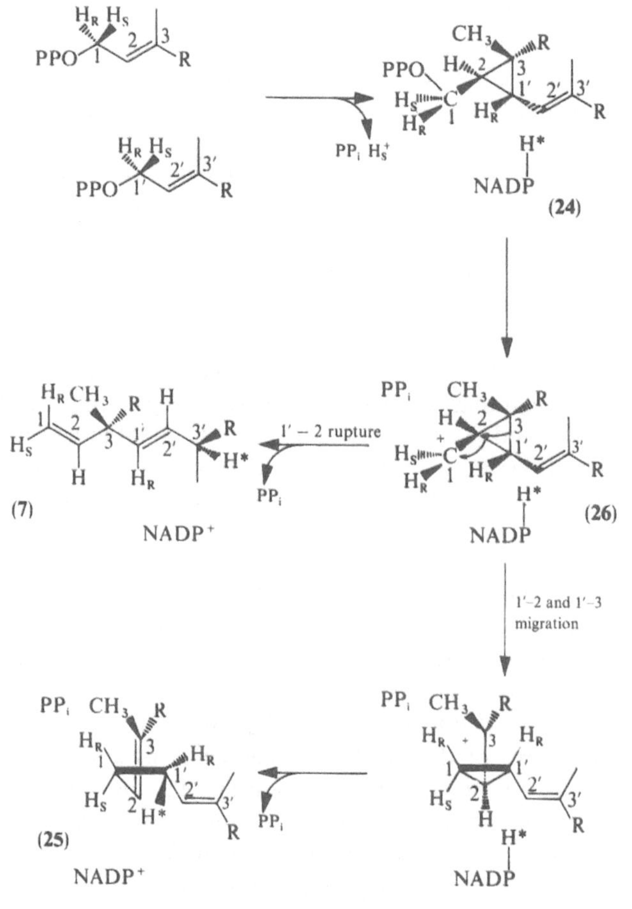

Fig. 7. A mechanism for biosynthesis of botryococcene (7) and squalene (25) from pre-squalene pyrophosphate (24) (R = $C_{11}H_{19}$). Reproduced with the permission of J. Amer. Chem. Soc. [**111**, 2713, (1989)]

in the B race of *B. braunii*, the botryococcene synthetase might be derived from squalene synthetase by a mutation which would be at the origin of the most prolific producer of non-head-to-tail terpenes.

1.2.2.2 Alkylation

Indications of the precursor role of the C_{30} botryococcene (7) towards its higher homologues were obtained by feeding ^{14}C labelled leucine, a tracer efficiently incorporated by *B. braunii* (47). While the total radioactivity of the external hydrocarbon fraction remained stable after a culture period of one day, a flux of radioactivity was observed from C_{30} (7) towards the C_{31} (11) and the C_{32} (12) botryococcenes. Further evidence was obtained using pulse-chase experiment with $^{14}CO_2$ (48) and chase experiment with [1,2-^{14}C] acetate (23). Moreover, incorporation experiments with [Me-^{14}C] and [Me-^{13}C] methionine proved that this amino acid, probably *via* its S-adenosyl form, is the source of methyl groups during the sequential alkylation of the four trisubstituted double bonds initially present in (7), thus leading to the acyclic hydrocarbons C_{31} (11), C_{32} (12), C_{33} (14) and C_{34} (16) and (18) (23).

On the basis of the absolute configuration of (16) it could be deduced that in this C_{34} hydrocarbon the extra methyls are installed at the *si* face of the trisubstituted double bonds (38); however nothing is known about the stereochemistry of deprotonation in the carbocationic intermediates. It can only be said that the proton lost during alkylation generally arises from a vinylic methyl, with the exception of C_{34} (18) in which a proton was removed from the allylic methylene C(5), when methylation occurred at C(7), and of C_{33} (14) where deprotonation occurred on the methyl group introduced at C(3).

Fig. 8 shows a proposed biogenetic relationship between the C_{30}botryococcene (7), and the hypermethylated botryococcenes C_{36} (19) and C_{37} (20), involving the C_{34} compound (16) as intermediate. This tentative pathway, closely related to the triple and quadruple methylations of sterol side chain in sponges (40), would start with methylation of an α,α'-disubstituted double bond of C_{34} (16).

The structures of the cyclobotryococcenes suggested that methylation could be the starter of cyclization (32) in a way similar to that previously hypothesized for the formation of methylhopanoids (49) and of the precursors of the irones (42). A plausible mechanism for the biosynthesis of C_{34} (21) is outlined in Fig. 9. Methylation at C(20) could initiate cyclization followed by elimination of a proton so as to produce an olefin which is in turn methylated at C(18) to finally give the pentamethylated cyclohexenyl ring. Stereochemical studies on the cyclo-

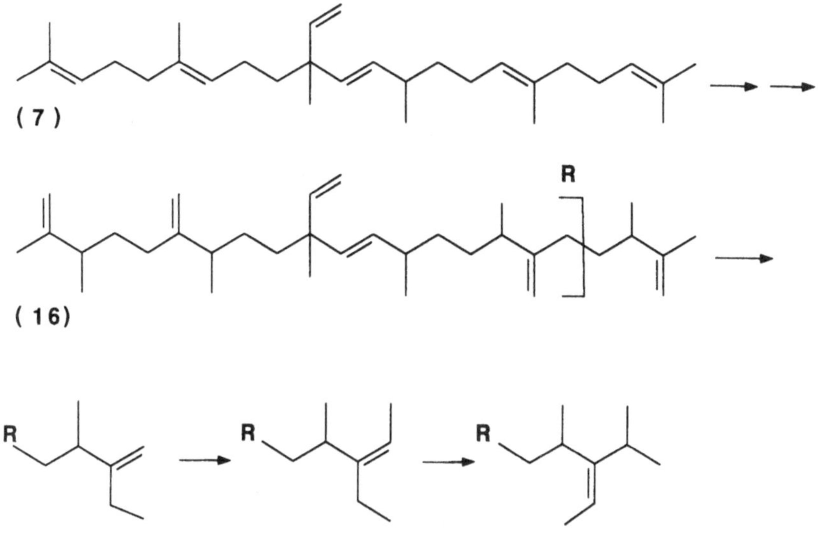

Fig. 8. Proposed biogenetic pathway of botryococcenes C_{36} (19) and C_{37} (20)

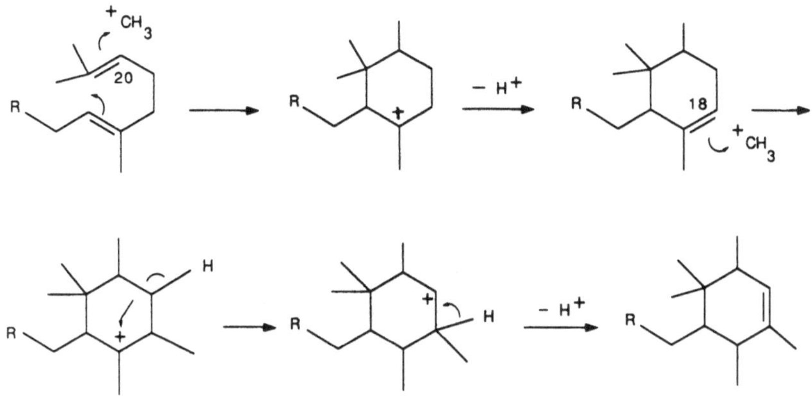

Fig. 9. Proposed mechanism for cyclization of C_{34} cyclobotryococcene (21) (R = $C_{22}H_{37}$)

botryococcene C_{32} (23), suggested another possible cyclization process. In this latter mechanism methylation would occur before ring closure, leading to the acyclic botryococcene C_{32} (12); cyclization of (12) would be initiated by protonation of the C(21)–C(22) α,α'-disubstituted double bond (43).

Finally, all these methylation patterns conjointly with the occurrence of four possible sites for methylation can explain the great number of botryococcenes produced by *B. braunii*. Poulter and Huang considered that this richness could be related to the existence either of a few isoprenoid transmethylases with rather low substrate specificities or of a large family of highly specific enzymes (*50*). The identification, in addition to the botryococcenes and squalene, of the monomethyl-squalene C_{31} (**27**) (*51*) and of the tetramethyl-squalene C_{34} (**28**) (*50*) in two different strains of *B. braunii*, supports the former assumption.

(27)

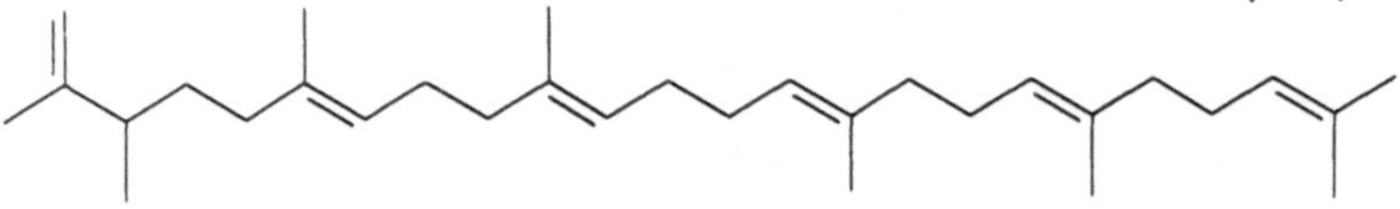

(28)

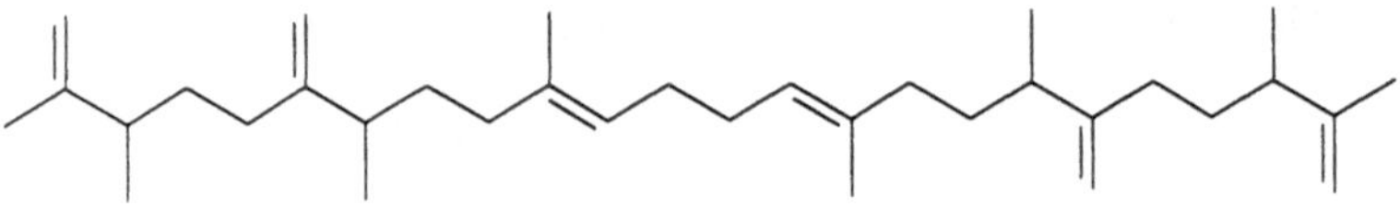

1.2.3 Chemical Variability and Abundance of Hydrocarbons in the B Race

The first comparative survey of botryococcene producing algae was carried out by Wake and Hillen who analyzed *B. braunii* samples collected from four Australian lakes (Devilbend, Darwin Reservoir, Green Lake and McCay's Reservoir) (*13*). GCMS analyses pointed to a large variation in botryococcene composition depending on sample origin. Moreover, for a given lake, the authors noted a smooth evolution with time in the relative abundance of these hydrocarbons. Similar results were also observed with algae collected from West Indian lakes (*11*). From analyses performed on laboratory grown strains (Table 3) it was shown that both genetic and physicochemical factors are responsible for botryococcene variability (*11*): (i) strains from different geographical origin cultivated under strictly identical conditions exhibit different hydrocarbon profiles (Table 3); (ii) when strains were cultivated under conditions promoting a fast growth, *i.e.* when CO_2 was supplied to the cultures, a change in hydrocarbon composition was observed over time: lower botryococcenes were produced at first, while higher compounds

Table 3. *Distribution (%) and Content of External Botryococcenes in a Wild Sample* and in Two Cultivated Strains of the B Race*

C_nH_{2n-10} n =	Structures	Origin Australia Darwin*	Bolivia Overjuyo (strain 5)	West Indies Martinique-La Manzo (MLM1 strain)
32	(12)	–	10.4	10.9
32	(23)		18.7	–
33	(14)	–	3.9	85.9
34		8.3	6.6	–
34	(16)	19.0	54.9	3.1
34		2.4	–	–
34	(21)	15.2	–	–
34		3.5	–	–
35		2.6	–	–
36	(19)	34.8	–	–
37	(20)	10.3	–	–
Others		3.9	3.8	–
Content†		29.2	not determined	32.0

Data taken from refs. *11* and *17*.
† Total botryococcenes as % of dry wt.

accumulated in older cultures. This result can be explained on the one hand by differences in the relative rates of biosynthesis of the precursor of all botryococcenes, the C_{30} (7), and of the methylation processes, and on the other hand by the precursor role played by each botryococcene towards its next higher homologue. Finally in old cultures, the lower compounds disappeared to the benefit of one or several end botryo-coccenes. Because of these variations during growth the hydrocarbon profiles of different strains must be established on algae exhibiting the same physiological development and cultivated under identical conditions.

Morever, marked differences in botryococcene distribution (large predominance of C_{33} or C_{34}) were noted in two strains isolated from a West Indian lake, whereas the above compounds were found in almost equal amounts in the mother sample collected from the lake (*11*). Distinct populations of botryococcene producing algae can therefore coexist in nature.

Finally, in the B race, variations in hydrocarbon content in relation with the strain origin occur to a lesser extent when compared with the A race; high levels of botryococcenes ranging from 25% to 45% dry wt are currently observed in wild samples or cultivated strains (*11, 13*).

1.2.4 Botryococcene Synthesis

The key in botryococcene synthesis lies in the construction of the central unit occurring in all the compounds of the series. This unit comprises the C(10) quaternary center, the *trans* C(11)–C(12) double bond and the C(13) methyl group. Two strategies were developed to

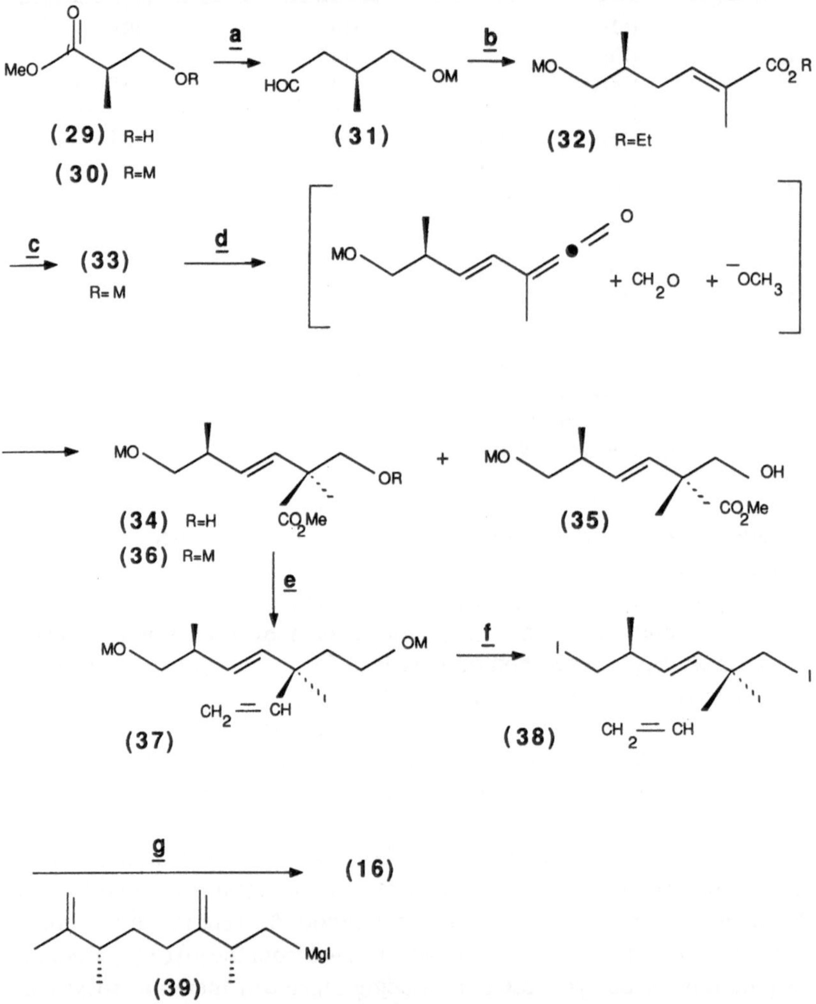

Fig. 10. Synthesis of (−) C_{34} botryococcane (**16**). *a* [1: LiAlH$_4$, 2: p-TsCl, 3: KCN, 4: (i-Bu)$_2$AlH]; *b* [Ph$_3$PC(CH$_3$)CO$_2$Et]; *c* [1: KOH, 2: ClCH$_2$OCH$_3$]; *d* [(i-Pr)$_3$NLi-HMPA]; *e* [1: LiAlH$_4$, 2: (COCl)$_2$, 3: Ph$_3$PCH$_3$]; *f* [1: HCl, 2: p-TsCl, 3: NaI]; *g* [1: CuI, 2: (**39**)]. M: methoxymethyl group

produce this central unit, both using (*R*)-methyl 3-hydroxy-2-methyl-propionate (**29**) as starting material.

In the synthesis depicted in Fig. 10, WHITE *et al.* (*52*) converted the protected methoxymethyl (MOM) ether (**30**) derived from (**29**) into the aldehyde (**31**), followed by a Wittig reaction to obtain the *E* α,β-unsaturated ester (**32**). Subsequent reaction of the bis-MOM derivative (**33**) with lithium diisopropylamide furnished, *via* a fragmentation-recombination process, two diastereoisomers (**34**) and (**35**). After chromatographic separation of (**34**), its bis-ether derivative (**36**) was converted into

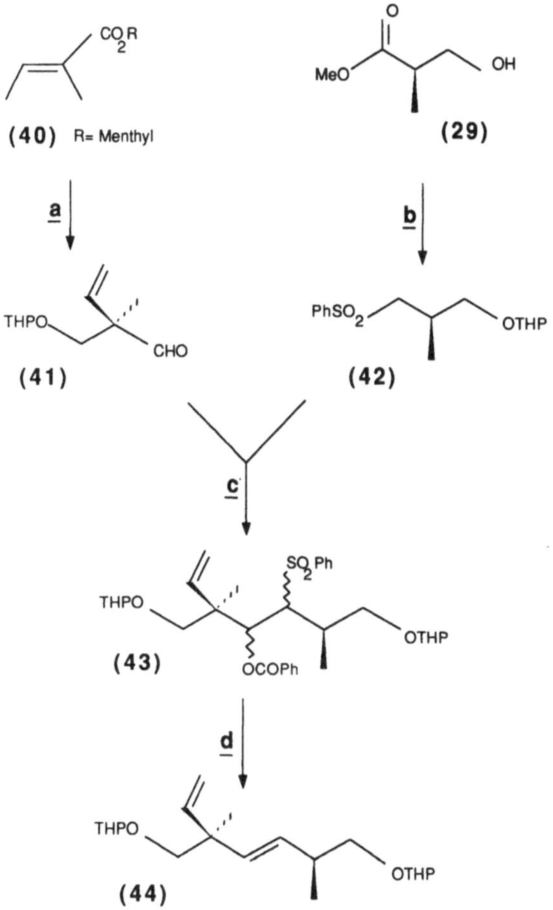

Fig. 11. Synthesis of the central unit of botryococcenes. *a* [1: (i-Pr)$_2$NLi, 2: HCHO, 3: dihydropyran/H$^+$, 4: LiAlH$_4$]; *b* [1: dihydropyran/H$^+$, 2: LiAlH$_4$, 3: p-TsCl and NaI, 4: PhSO$_2$Na]; *c* [*n*-BuLi, PhCOCl]; *d* Na(Hg)

an aldehyde which was transformed to (37) in a second Wittig reaction. Finally, the naturally occurring (−)-botryococcene C_{34} (16) was obtained by condensing the diiodide (38) with the Grignard reagent (39) prepared in fourteen steps from the (S)-enantiomer of (29).

More recently, HIRD et al. (53) used another method, depicted in Fig. 11, for preparation of the central unit of the botryococcenes. The synthesis was based on the coupling of the quaternary center (41), prepared from menthyl tiglate (40), with the sulphone (42) derived from (29). Reduction of the vicinal benzyloxysulphone (43) with sodium amalgam afforded the central building block (44). In parallel work, the authors prepared the 10-(R, S) botryococcene C_{30} and the corresponding botryococcane using a similar strategy.

1.3 Lycopadiene (L Race)

This compound was isolated from three strains of B. braunii originating from tropical freshwater lakes (Table 1) as their sole hydrocarbon (19, 20); it accounted for 2–10% of dry wt, depending on the strain origin and the culture age. The EIMS indicating a molecular ion at m/z 558 and the elemental analysis supported the formula $C_{40}H_{78}$. The ^1H and ^{13}C NMR spectra agreed with a terpenoid structure. Thus the ^1H NMR spectrum showed signals for two vinylic methyls at $\delta 1.58$, four aliphatic methyls at $\delta 0.86$ and four others at $\delta 0.84$; a triplet at $\delta 5.13$ was indicative of two trisubstituted double bonds. A trans configuration was assigned to these unsaturations on the basis of the chemical shifts observed for the allylic carbons in the ^{13}C NMR spectrum.

Hydrogenation in the presence of Rh/C furnished a $C_{40}H_{82}$ hydrocarbon exhibiting the same mass spectrum and GC retention time as lycopane, the entirely reduced derivative of the acyclic C_{40} carotenoids.

Ozonolysis of lycopadiene followed by oxidative cleavage of the ozonide afforded 6,10,14-trimethylpentadecan-2-one thus establishing the C(14)–C(15) and C(18)–C(19) location of the unsaturations. Moreover, the optical rotation of this ketone was identical with that observed for the 6(R),10(R), 14-trimethylpentadecan-2-one obtained from ozonolysis of natural phytol; this result allowed assignment of stereochemistry to the asymmetric centers. Therefore, lycopadiene (45) was identified as

(45)

2,6(R), 10(R), 14,19,23(R), 27(R), 31-octamethyldotriaconta-14(E), 18(E) diene.

On the basis of the configuration of the four chiral centers, a biosynthetic pathway can be proposed with phytol as precursor. In this mechanism, two phytyl chains would condense, probably as pyrophosphate derivatives, to produce the prelycopadiene pyrophosphate (46), which would be converted to lycopadiene (45) by addition of a hydride ion to a rearranged cyclopropylcation. An intermediate analogous to (46) is known to occur in the synthesis of phytoene, the precursor of the carotenoids; in this latter case two geranylgeranyl pyrophosphates are condensed (45). In another possible pathway, lycopadiene (45) would merely result from reduction of phytoene.

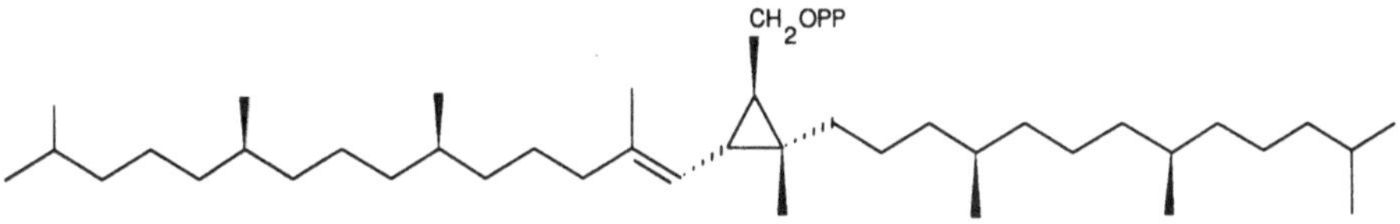

(46)

2. Classical Lipids

2.1. n-Fatty Acids

Even-carbon-numbered *n*-fatty acids, in the range C_{14}–C_{30} are important constituents of the three races of *B. braunii* where they account for 5 to 14% of dry biomass depending on the strain (20); these are obtained by saponification of hexane and CHCl$_3$-MeOH extracts. For a given race, GCMS analyses of their methyl esters indicated rather similar distribution in external and internal lipids. The most abundant fatty acids are palmitic (16:0), oleic (18:1) and octacosenoic (28:1) acids. Oleic acid is markedly predominant in the A race (more than 80% of total fatty acids); this very high level is certainly related to the precursor role of this acid in *n*-alkadiene and triene biosynthesis. The above *n*-fatty acids are esterified in the form of various lipid classes including triacylglycerols (14).

2.2 Triacylglycerols

Triacylglycerols isolated from the hexane extracts account for 2 to 6% of dry biomass in the A race (14). Analysis by reversed-phase HPLC on a C_{18} column showed that triolein is the major constituent of this fraction and that some molecular species contain acyl moieties exhibiting very long mono-unsaturated chains up to C_{28}, so far only detected in triacylglycerols from higher plants.

2.3 Sterols

The free sterol fraction is composed of cholest-5-en-3-β-ol and of its 24-methyl and 24-ethyl derivatives in the three races accounting together for 0.1 to 0.2% of the dry biomass, they were recovered primarily from the internal lipids (20). While often considered as good taxonomic markers, sterols do not permit a distinction between the different populations of B. braunii. Some other minor sterols still remain to be identified in the L race.

2.4 Carotenoids

Carotenoids were studied in the B and L races where they account for 0.2 to 0.4% of dry biomass. Ten carotenoids were identified on the basis of their chromatographic properties (TLC, HPLC on nitrile and silica columns), visible, ^{1}H-NMR and CD spectra and comparison with authentic compounds (54). Both races produce 6(R)-β,α-carotene (47), β,β-carotene (48), echinenone (49), (3'R)-3-hydroxyechinenone (50), canthaxanthin (51), (3R,3'R,6'R)-lutein (52), (3R,3'R)-zeaxanthin (53), (3S,5R,6S,3'S,5'R,6'S)-violaxanthin (54), (3R,3'R,6'R)-loroxanthin (55) and (3S,5R,6R,3'S,5'R,6'S)-neoxanthin (56). The most significant difference between the B and L races lies in the higher content of (50) in the former.

In the stationary phase of growth characterized by an intense orange to red colour, echinenone (49) and canthaxanthin (51) are major carotenoids; they are preferentially located in the outer walls, while other carotenoids are intracellular (55). Recent analyses of N-deficient and N-rich cultures demonstrated that initiation of the synthesis of canthaxanthin (51) is concomitant with the onset of N-deficiency (55). On structural grounds, β,β-carotene (48) and echinenone (49) are considered as the biosynthetic precursors of (51).

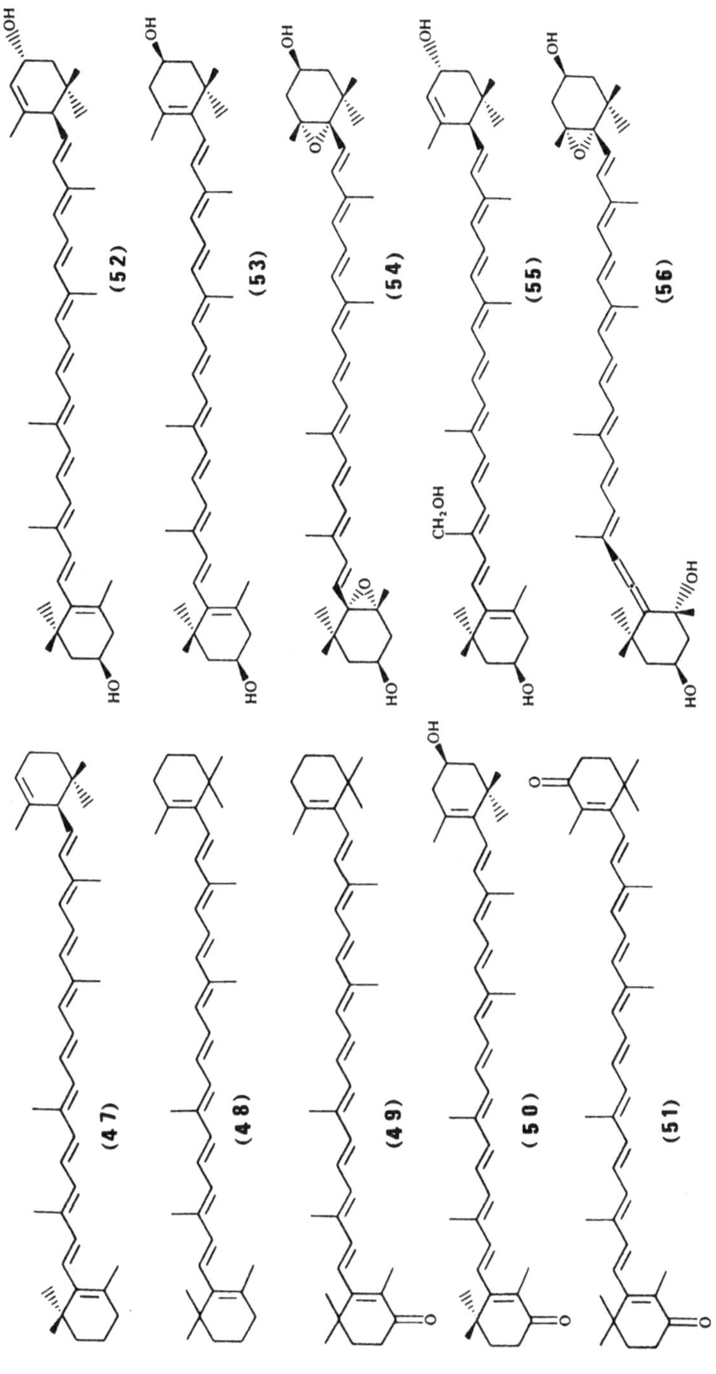
(52)
(53)
(54)
(55)
(56)
(47)
(48)
(49)
(50)
(51)

3. Non-classical Lipids of the A Race

3.1 Botryals

The botryals, a family of even carbon numbered C_{52}–C_{64} α-branched, α-unsaturated aldehydes (56) have been detected in all the strains of the A race so far investigated, but their abundance shows marked variations with strain origin. Thus the botryals are the most important compound class (*ca* 45%) in the external lipids of algae originating from Lake Titicaca, while they are minor products (2.6% of external lipids) in a French strain isolated from Lake Coat ar Herno (14).

This family comprises two series of compounds: botryals (57) and (58) obtained after separation by preparative TLC. These isomeric series differ only in the Z and E configuration respectively of the double bond conjugated with the carbonyl group. The IR spectra of (57) and (58) showed an absorption for CH of an aldehyde at 2720 cm^{-1}, associated with a C=O stretch shifted at 1680 cm^{-1} for (57) and 1690 cm^{-1} for (58) by the conjugated unsaturation. ^{1}H- and ^{13}C-NMR spectra supported the presence of one α-unsaturated and α-branched aldehydic function, two isolated double bonds and two terminal methyls. Configurations of unsaturations were assigned by ^{13}C-NMR spectrometry: (i) the allylic carbons of the two isolated double bonds in the chains resonate at 27.3 both in (57) and (58) as usually observed for *cis* stereochemistry; (ii) the signal of the carbonyl carbon is more deshielded in (58) (δ195.0) than in (57) (δ190.9), thus indicating that the formyl group was *trans* and *cis* relative to the β-alkenyl chain respectively. Finally, ozonolysis allowed determination of the position of the double bonds.

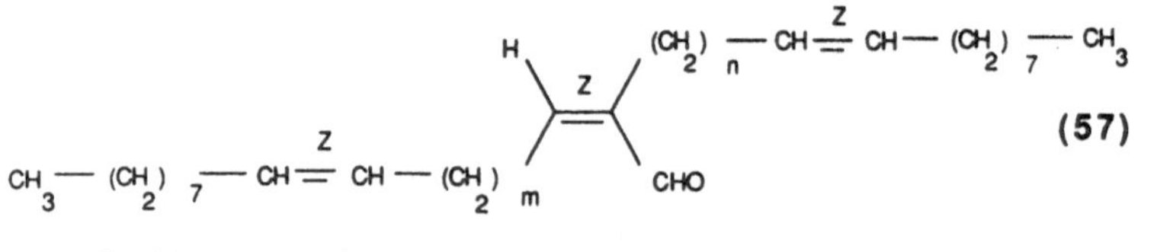

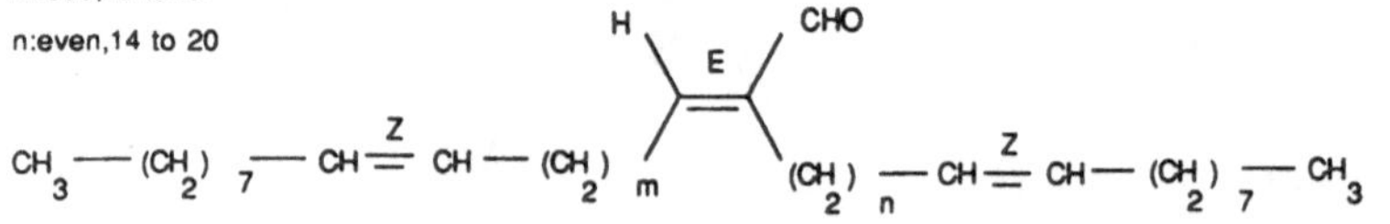

Regarding botryal distribution, the two series of isomers (57) and (58) were not resolved into individual compounds by normal- and reversed-phase HPLC and did not elute on GC columns. Yet, based on the relative intensities of the M^+ ions $[C_nH_{2n-4}O]^+$ observed in the probe EI mass spectra, with $52 \leq n$ (even) ≤ 64, a predominance of the C_{58} compounds was noted for the two series.

In the course of the botryal study, it was observed that such compounds undergo a rather fast transformation into hydrocarbons of a very high molecular weight. Thus, series (**58**) kept in air at room temperature led, after one week, to a series of *n*-alkatrienes (**59**) (8% of the starting botryals) of general formula $C_{51+x}H_{98+2x}$, x even from 0 to 12. These trienes, with two *cis* ω9 and ω9′ double bonds and a centrally located *trans* unsaturation, presumably arose from spontaneous decarboxylation of the botryal-derived acids; concomitantly isomerisation of the double bond initially conjugated with the C=O (of *cis* configuration with respect to the alkenyl chains) would also occur. The trienes (**59**) were not detected in living algae.

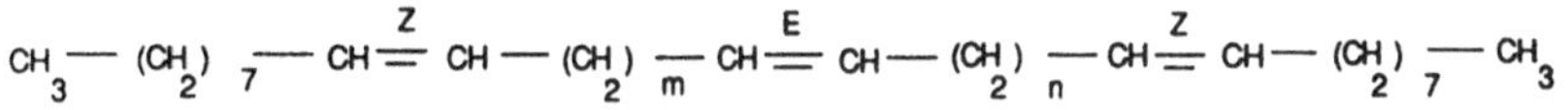

$$CH_3-(CH_2)_7 \overset{Z}{-CH=CH}-(CH_2)_m \overset{E}{-CH=CH}-(CH_2)_n \overset{Z}{-CH=CH}-(CH_2)_7-CH_3$$

m and n as in (**57**) and (**58**)   (**59**)

Structures (**57**) and (**58**) suggested two possible origins for the branched CHO occurring in the central unit: either a formylation reaction or a head-to-head condensation mechanism similar to that previously envisaged and now discarded for alkadiene biosynthesis. Feeding experiments with sodium $[1,2-^{13}C]$ acetate provided evidence for the latter mechanism. Indeed, the ^{13}C-NMR spectra of ^{13}C enriched (**57**) and (**58**) established the repeated condensation of C_2 units, probably via malonyl CoA, leading to the even, C_{26} to C_{32}, very long chain *n*-fatty aldehydes $R^1-^{13}CH_2-^{13}CHO$ and $R^2-^{13}CH_2-^{13}CHO$. Afterwards, these aldehydes would be condensed and the resulting aldols dehydrated, as depicted in Fig. 12. Another argument for such a mechanism lies in the

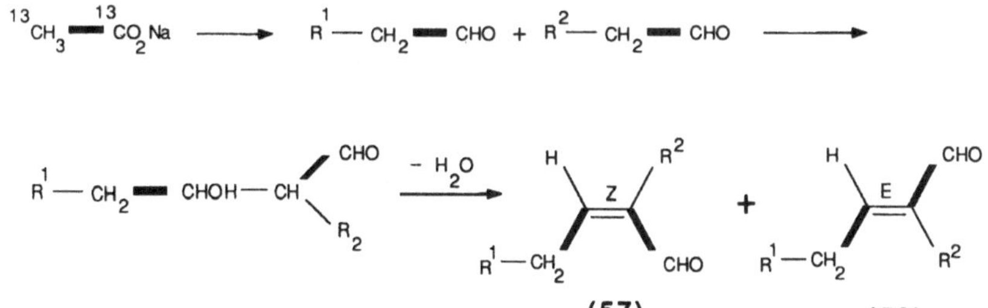

Fig. 12. Proposed biosynthesis of botryals from $[1,2-^{13}C]$ acetate

higher amounts of botryals (**58**) in all the tested strains. Indeed, this predominance of the isomers where the formyl group and the alkyl chain are *trans* is also observed for acid- or base-catalysed dehydration of aldols.

From a structural and a biosynthetic point of view, the botryals therefore appear closely related to mycolic acids, important lipid constituents occurring in the cell walls of *Corynebacterium, Nocardia* and *Mycobacterium*; these compounds are α-branched, β-hydroxy acids synthesized by head-to-head condensation of long chain fatty acids (*57*).

3.2 n-Alkenylphenols

Very long chain *n*-alkenylphenols (**60**) accounting for *ca* 1.5% of dry biomass were identified in the external lipids of the collection strain from Austin (*56*) and detected in all the strains of the A race examined so far. Isolated by TLC as a solid mixture, these compounds exhibit an UV absorption maximum at 215 nm (ε 7500) and IR bands at 1600 and 3560 cm^{-1} both indicative of a phenol moiety.

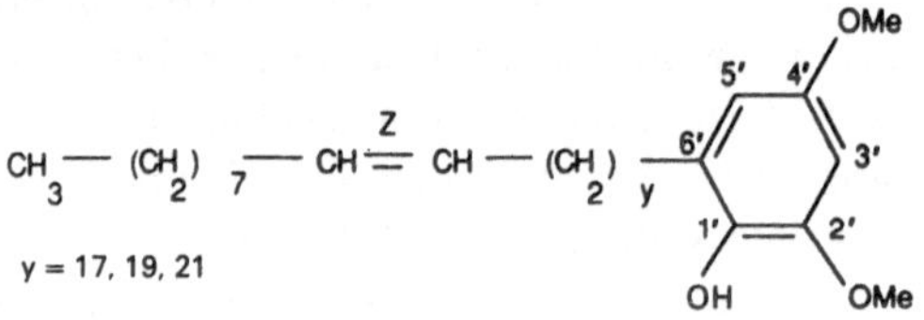

$$CH_3 - (CH_2)_7 - CH \overset{Z}{=} CH - (CH_2)_y - \text{(ring)}$$

y = 17, 19, 21   (**60**)

The probe EI and CI(NH$_3$) mass spectra showed the presence of three compounds: $C_{35}H_{62}O_3$, $C_{37}H_{66}O_3$ and $C_{39}H_{70}O_3$ in approximately 3:12:5 ratio. The EI mass spectrum, which exhibited a peak at m/z 167 resulting from a benzylic cleavage and the ^{1}H- and ^{13}C-NMR spectra established the presence of a hydroxydimethoxyphenyl unit connected to a long *n*-alkenyl chain. Two doublets in the ^{1}H-NMR spectrum at δ 6.29 and 6.39 with a small coupling constant (J = 2.8 Hz) were characteristic of a tetrasubstituted phenyl group with two *meta*-orientated hydrogen atoms. Comparison of the ^{13}C-NMR chemical shifts of the aromatic carbons in compounds (**60**) with those of trimethyl-ether derivatives allowed assignment of the relative position of OH and OMe groups on the benzene ring. The position of the "in chain" unsaturation was determined by ozonolysis and its configuration from ^{13}C-NMR data.

A polyketide origin for (**60**) was established by feeding experiments with sodium [1,2-^{13}C] acetate (Fig. 13). According to the coupling constant (43 Hz) between C(6′) and the benzylic carbon, these two carbons derive from the same acetate unit condensed to form a very long chain fatty acyl derivative. In addition, the presence of pairs of satellites around the peaks of C(2′), C(3′) and C(4′) and of satellites around those of C(1′) and C(5′), these two latter central peaks exhibiting also substantially higher intensity, suggested the formation of two non-equivalent ^{13}C-enriched trioxygenated derivatives (**61**) and (**62**). The hydroxyl group at (C1′) would result from oxidation at one of the two ortho positions of a symmetrical intermediate (**63**).

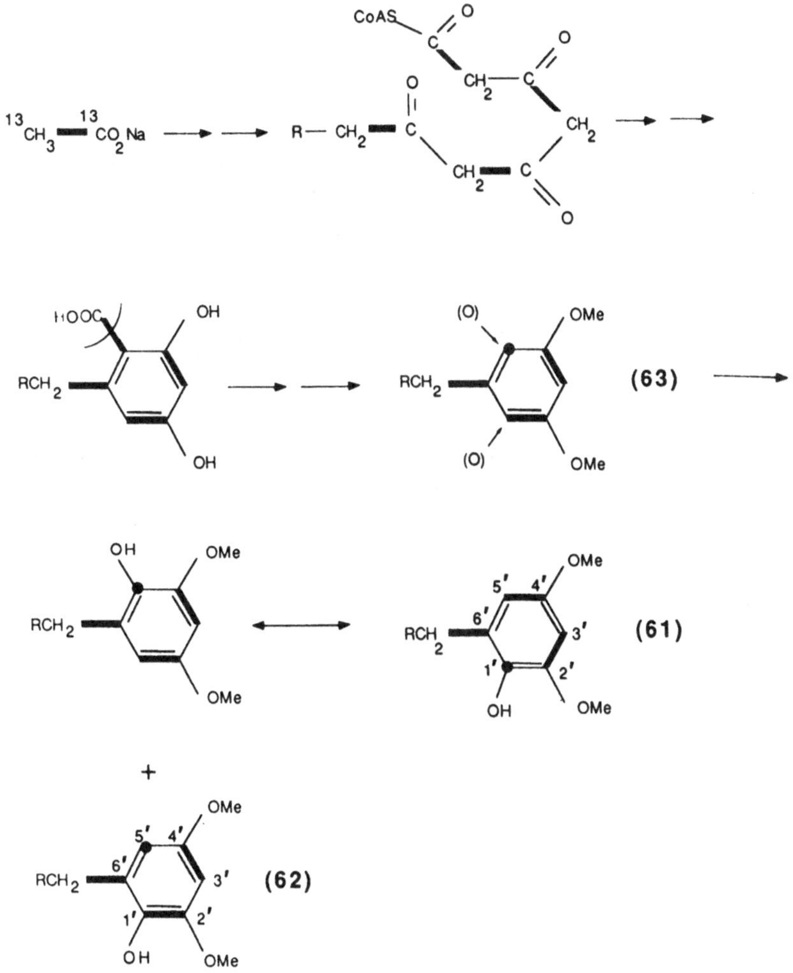

Fig. 13. Proposed biosynthesis of *n*-alkenylphenols from [1,2-^{13}C] acetate

The alkyl- and alkenylphenols occurring in terrestrial plants are known to exhibit antifungal activity (58) and the major role of their hydrocarbon chain might be to increase the solubility of the phenol moiety in lipidic region of plants which require protection against microbial or oxidative degradations. In a similar manner, the alkenylphenols of the A race might protect the aliphatic compounds, essentially unbranched or poorly branched, located in the outer walls against degradation by bacteria and fungi. In sharp contrast, the B and L races which are rich in terpenoids, compounds known for their higher resistance towards microbial attacks, do not synthesize this type of phenols.

3.3 Epoxides

Three series of epoxides, accounting together for *ca* 4% of dry biomass were isolated by TLC from the external lipids of the Austin collection strain (56). EIMS, ^{1}H- and ^{13}C-NMR data showed that these were epoxyalkenes (64), epoxybotryals (65) and epoxyalkylphenols (66) derived from the parent compounds (1), (58) and (60) respectively, by epoxidation of the C(ω9)–C(ω10) double bond. All these compounds contain a *cis* epoxide ring as deduced from the ^{13}C chemical shifts of the α-methylene carbons. In the case of the epoxybotryals (65) the location of the epoxide relatively to the two hydrocarbon branches was not ascertained and two isomeric series might occur. Neither the diepoxyalkanes, nor the 1,2-epoxyalkenes derived from *n*-alkadienes (1) were detected in the lipids of the Austin strain.

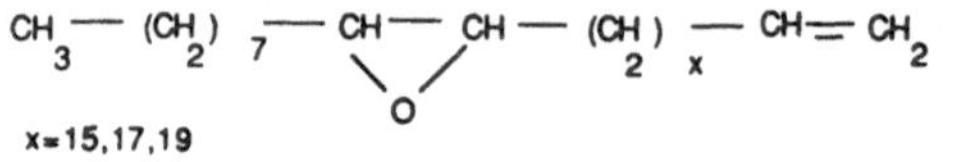

$$CH_3 \!-\! (CH_2)_7 \!-\! CH \!-\! CH \!-\! (CH_2)_x \!-\! CH \!=\! CH_2 \quad \textbf{(64)}$$

x = 15, 17, 19

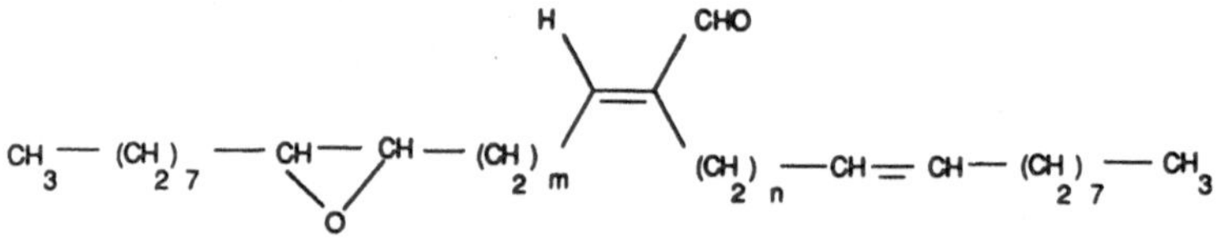

m and n as in (57) and (58)   (65)

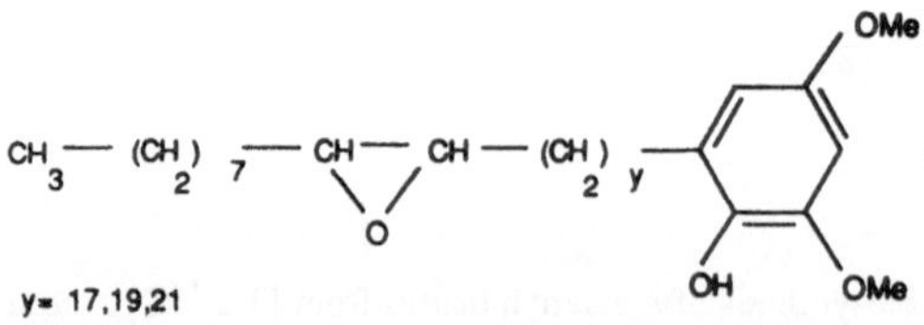

y = 17, 19, 21   (66)

Given the diversity of these epoxide families, it may be assumed that the epoxidase implicated in the oxidation of the double bonds has broad substrate specificity regarding remote substitutions (beyond fifteen carbons), occurring in the hydrocarbon chain. At first, the richness of the A race in epoxides was surprising; however as discussed below, these epoxides could function as precursors in the biosynthesis of molecules of a more complex structure.

3.4 Ether Lipids

The first compounds of this group, the alkatrienyl-alkadienyl ether (**67**) and the alkyl-alkadienyl ethers (**68, 69**) were discovered in two strains of the A race originating from the Bolivian Lake Overjuyo, and the French Lake Coat ar Herno (*59*). In laboratory cultures these algae produced impressive levels of ether lipids, as high as 27% of dry biomass, while hydrocarbons were only minor products (2% of dry biomass). Generally ether lipids occur as glycerol derivatives in biological membranes, for example 1-O-*n*-alkyl or 1-O-*n*-alkenyl glycerols in animal tissues, or in few cases as divinyl ether fatty acids arising from the action of a lipoxygenase such as in potato tubers. However, in sharp contrast to the ether lipids in *B. braunii*, such compounds never predominate.

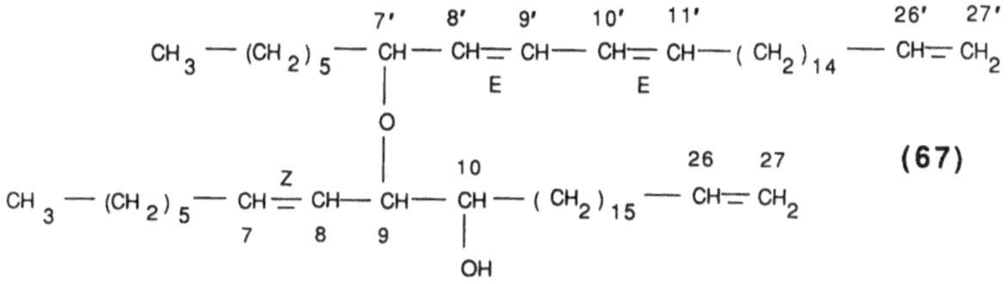

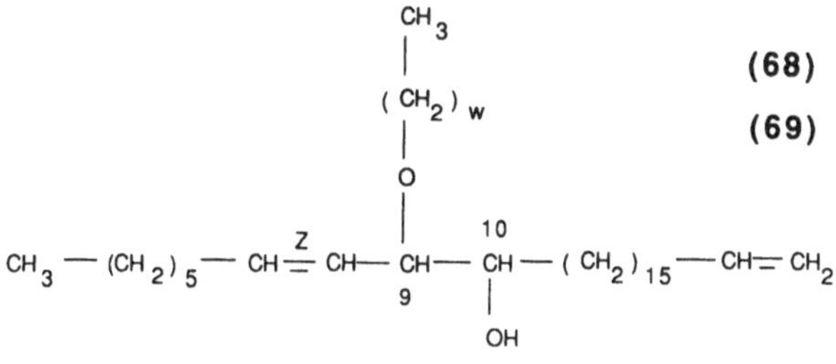

Compounds (**67**)–(**69**) termed botryococcoid ethers were isolated from external lipids and purified by TLC and HPLC. The EI and

CI(NH$_3$) mass spectra of the most abundant one, the alkatrienyl-alkadienyl ether (67) (accounting alone for up to 26% of dry biomass), established the formula C$_{54}$H$_{100}$O$_2$. From IR, ^{1}H- and ^{13}C-NMR spectra, it was shown that (67) contains one secondary hydroxyl group and an ether function binding two methine carbons. Hydrogenolysis of (67) on Pd/C gave in equal amounts n-heptacosane and a mixture of n-heptacosan-7-ol and n-heptacosan-10-ol with the former isomer predominating, thus indicating that two n-C$_{27}$ chains occur in the ether lipid. Extensive proton decoupling and a COSY experiment revealed the presence of (i) two terminal double bonds; (ii) a four-proton unit where a methine of the ether function (δ 4.19) was coupled both with the proton on the hydroxyl-bearing carbon (δ 3.68) and with an olefinic proton (δ 5.29), the latter in turn coupled with a cis olefinic proton (δ 5.62, J = 11 Hz); (iii) another five-proton unit where the second methine proton of the ether function (δ 3.92) was α to a trans double bond (δ 5.43 and 6.20, J = 14.9 Hz), conjugated with a second trans unsaturation (δ 6.06 and 5.67, J = 14.7 Hz).

The position of the functions of the two C$_{27}$ chains was determined both from the major spectral fragments observed in the EIMS as shown in Fig. 14, and by identification of n-C$_{16}$ and n-C$_{17}$ diacids in ozonolysis by-products. All these results led to establishment of structure (67) for the alkatrienyl-alkadienyl ether.

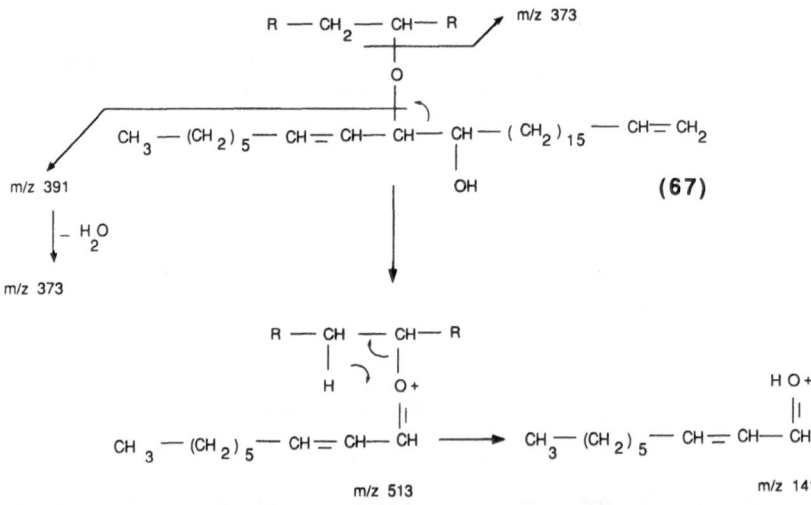

Fig. 14. Major spectral fragments of botryococcoid ether (67)

The alkyl-alkadienyl ethers (**68**) and (**69**) were isolated as a white solid mixture; these minor compounds from the Overjuyo and Coat ar Herno strains (together amounting to 1.1% of dry biomass) were identified as $C_{51}H_{100}O_2$ and $C_{53}H_{104}O_2$ respectively. ^{1}H, ^{13}C-NMR and MS spectra indicated that (**68**) and (**69**) contain the same hydroxyheptacosadienyl chain as (**67**), but here the oxygen of the ether is bound to a methylene carbon of a saturated straight hydrocarbon chain: $C_{24}H_{49}$ in (**68**) and $C_{26}H_{53}$ in (**69**).

The general structure of the botryococcoid ethers suggested a close biosynthetic relationship with some lower metabolites occurring in the above strains such as the n-C_{27} triene (**4**) which is predominant in their hydrocarbon fraction. Thus, when chains are compared in (**4**) and (**67**), several common features are noted, including the same number of carbon atoms, the occurrence of a terminal double bond and the presence of unsaturation(s) in the C(7) to C(10) positions. Moreover, a precursor-product relationship was observed between the n-C_{27} triene (**4**) and (**67**), the level of the triene decreasing when the production of (**67**) strongly increased (*25*). Feeding experiments with a mixture of ^{14}C-labelled hydrocarbons, including the triene (**4**), obtained *via* incubation of the algae with sodium [1,2-^{14}C] acetate, confirmed this assumption since a substantial amount of radioactivity was incorporated into (**67**) without prior degradation of the fed hydrocarbons.

Despite the lack of direct evidence, the most likely mechanism for the formation of botryococcoid ether (**67**) is the coupling of two mono-epoxides (**70**) and (**71**) derived from the trienic hydrocarbon (**4**) (Fig. 15).

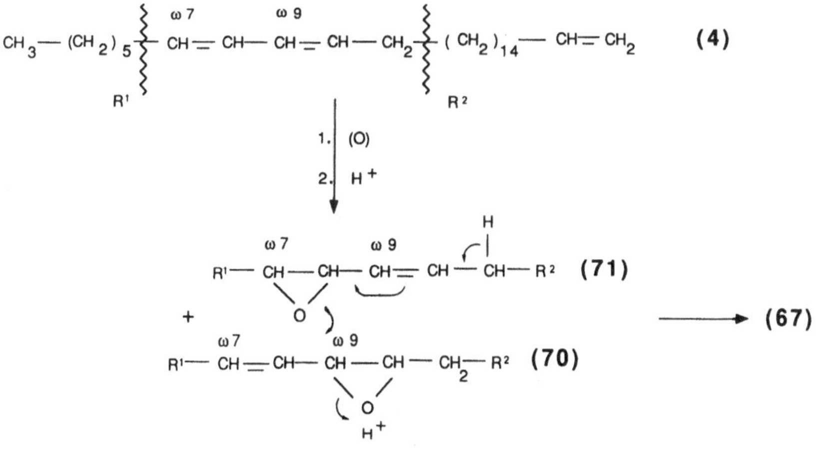

Fig. 15. Proposed biosynthesis of botryococcoid ether (**67**) from *n*-alkatriene (**4**)

The involvement of epoxides as intermediates in ether formation was previously suggested, for example, in the biosynthesis of some polyether toxins in dinoflagellates (60). The coupling reaction might be initiated by protonation of the epoxide group in (70), giving rise after ring opening to a positive charge at C(9). In a second step, this carbocation would react with epoxide (71) to form the ether bridge; the associated migration of the $\Delta(9')$-unsaturation and abstraction of an allylic proton at C(11') would finally result in the *trans-trans* conjugated system of (67). The origin of the n-C_{24} and n-C_{26} saturated chains in (68) and (69), is less clear; they could derive from tetra- and hexacosanoic acids, identified among the n-fatty acids of the A race (20).

So far, the dioxygenated botryococcoid ethers (67–69) or homologous compounds have not been identified in strains of the A race which synthesize high amounts of hydrocarbons. The strain from the Austin collection, however, produces another group of ethers closely related to n-alkadienes (1), botryals (58) and alkenylphenols (60) (61). The simplest compounds of this second group of botryococcoid ethers, isolated from external lipids, are the dialkenyl ethers (72) formed by an ether linkage of two hydrocarbon chains, likely derived from n-alkadienes (1), and hydroxylated in α and α' of the ether function. Such compounds account for *ca* 0.3% of dry biomass in the Austin strain. Monoester derivatives (73) of these dialkenyl ethers were also identified but in higher amount (2% of dry biomass). Other minor constituents from the Austin strain are the alkylphenolalkenyl ethers (74) (0.2% of dry biomass) and the botryalalkenyl ethers (75) (1% of dry biomass).

EI and $CI(NH_3)$ mass spectra established that (72), (73) and (74) correspond to homologous series of compounds of high molecular weight differing by 28 uma: 802 to 914 for (72), 1066 to 1346 for (73) and 1248–1332 for (74). Since all MS techniques tested failed to produce the molecular peak, the botryalalkenyl ethers (75) were examined by size exclusion HPLC on an ultraStyragel column using THF as eluent; the retention volume was consistent with a M_w of 1650 $\pm$ 140 daltons.

The NMR spectra of ethers (72)–(75) contain signals characteristic of the parent compounds. COSY experiments established that in (73)–(75) the methine protons of the ether function resonated at δ 3.23 for H(10) and 3.43 for H(9'); the signal of proton H(10'), on the hydroxyl-bearing carbon, appeared at δ 3.49 and the signal of proton H(9), attached to the ester-bearing carbon, was at δ 5.02. In structure (72), the protons were assigned on the basis of spectral comparisons with the diacetate derivative: protons H(9) and H(10') on hydroxyl bearing carbons resonated at δ 3.57, whereas those attached to the carbons of the ether function, H(10) and H(9') were found at δ 3.32. The ^{13}C-NMR spectra of (73)–(75)

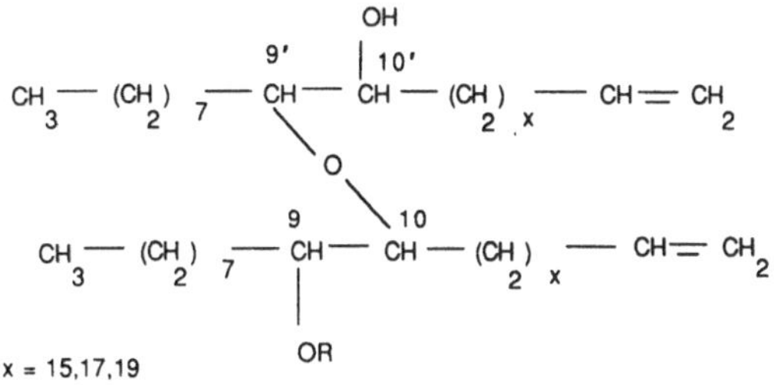

OH

9' |10'

CH₃ — (CH₂)₇ — CH — CH — (CH₂)ₓ — CH = CH₂   (72) R = H

O

(73) R = C₁₈ to C₃₀

9   10

CH₃ — (CH₂)₇ — CH — CH — (CH₂)ₓ — CH = CH₂   even fatty acyls

OR

x = 15,17,19

OMe

OH

CH₃ — (CH₂)₇ — CH — CH — (CH₂)ᵧ   (74) R as in (73)

O

HO   OMe

CH₃ — (CH₂)₇ — CH — CH — (CH₂)ₓ — CH = CH₂

OR

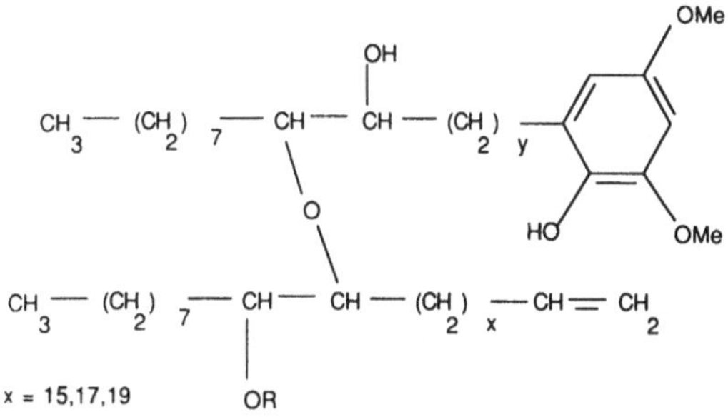

x = 15,17,19
y = 17,19,21

H   CHO

OH   ω10   ω9

CH₃ — (CH₂)₇ — CH — CH — (CH₂)ₘ   (CH₂)ₙ — CH = CH — (CH₂)₇ — CH₃

O

(75) R as in (73)

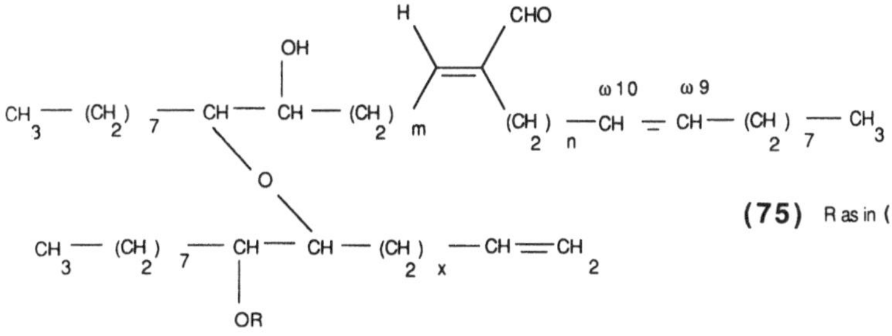

CH₃ — (CH₂)₇ — CH — CH — (CH₂)ₓ — CH = CH₂

OR

m = odd 15 to 21

x = 15,17,19

confirmed the existence in each series of the same hydroxyetherester pattern with four signals in the aliphatic ^{13}C-O region: δ C(9) 74.3, C(10) 77.5, C(9′) 80.9 and C(10′) 72.5.

In the EIMS, ethers (72)–(74) and their TMS derivatives exhibited similar fragmentation modes giving ions of substantial diagnostic value, as shown in Fig. 16 for the TMS derivatives of (72) and (73). The MS data demonstrated that in all these compounds the two hydrocarbon chains are linked through an ether bridge between C(9′) and C(10); the hydroxyl groups are located at C(9) and C(10′) in (72) whereas in (73)–(75) OH at C(9) is esterified by a fatty acid, essentially oleic acid.

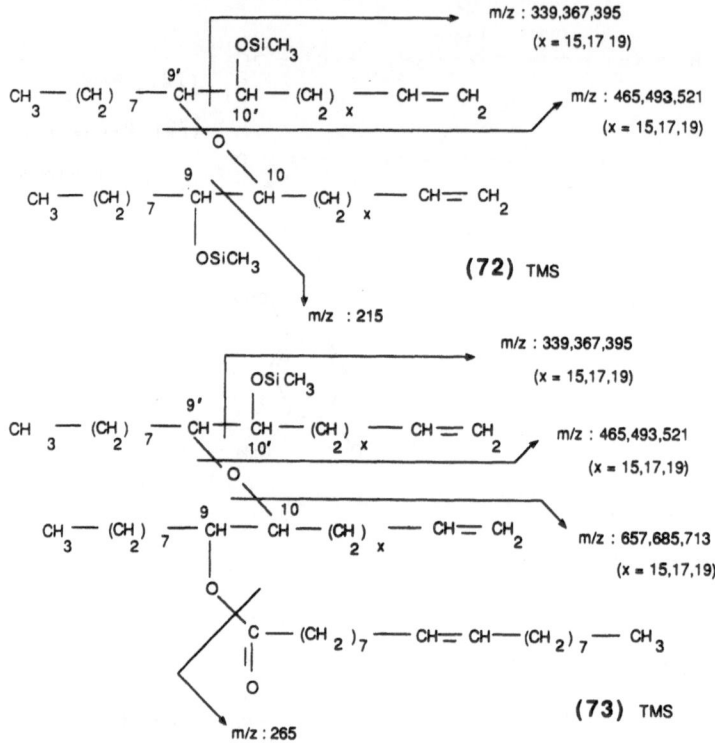

Fig. 16. Main EI-MS fragmentation patterns of (72) and (73) TMS derivatives

Again, epoxides could be close precursor in the biosynthesis of ethers (72)–(75). Coupling might start by protonation of an epoxide (64), (65) or (66) followed by the specific cleavage of the C(9′)–O bond; the carbocation formed would then attack a second epoxide (64) with concomitant cleavage of the C(9)–O bond. Then the positive charge at C(9) would be neutralized by reaction with water or a fatty acid, leading to (72) or (73)–(75), respectively.

From extensive TLC investigations of the external lipids of the Austin strain and analysis of the numerous fractions by IR and NMR spectroscopy, it was evident that several other ethers were also present; however attempts to isolate these compounds in pure form have been so far unsuccessful. These unresolved compounds are likely to be botryal-botryal ethers, botryal-alkylphenol ethers and isomers of (74) and (75) differing in the relative location of hydroxyl groups and ester in the two hydrocarbon chains. Moreover, the presence of an epoxide ring at C(ω9)–C(ω10), as shown by the NMR spectra, in some of the botryal moieties occurring in the ethers, would argue in favour of the formation of macromolecules comprising a larger number of cross-linked chains.

3.5 Methyl-branched Fatty Aldehydes and Fatty Acids

Methyl-branched aldehydes and acids, with odd numbers of carbon atoms from C_{23} to C_{29} were first discovered in the external lipids of the strain originating from Lake Coat ar Herno. The trimethyl fatty aldehydes (76), their α-unsaturated counterparts (77) and the trimethyl fatty acids (78) accounted for 2.7, 2.3 and 0.04% of dry biomass, respectively (62).

GC-EIMS of (76) showed the presence of an homologous series of four aldehydes $C_{23}H_{46}O$, $C_{25}H_{50}O$, $C_{27}H_{54}O$ and $C_{29}H_{58}O$ dominated by the C_{27} compound. The presence in each spectrum of a base peak at m/z 58 was indicative of a McLafferty rearrangement in an α-methylaldehyde. The NMR spectra confirmed the α-methyl branching and also established that two other methyl substitutions were present. IR and NMR spectroscopy revealed the presence of an α-unsaturated aldehyde function in (77); the configuration of the double bond was found to be *E* on the basis of ^{13}C-chemical shift comparisons with the two series of botryals (57) and (58). Selective hydrogenation of the carbon–carbon double bond in the presence of Pd/C afforded saturated aldehydes exhibiting the same spectral and chromatographic properties as (76). Trimethyl-fatty acids (78) were identified in the total fatty acid fraction by means of GC-EIMS. These odd C_{23}–C_{29} acids accompanying the

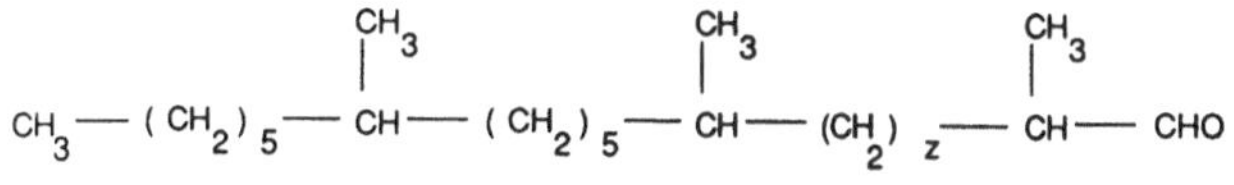

$$CH_3-(CH_2)_5-\underset{\underset{CH_3}{|}}{CH}-(CH_2)_5-\underset{\underset{CH_3}{|}}{CH}-(CH_2)_z-\underset{\underset{CH_3}{|}}{CH}-CHO \quad \textbf{(76)}$$

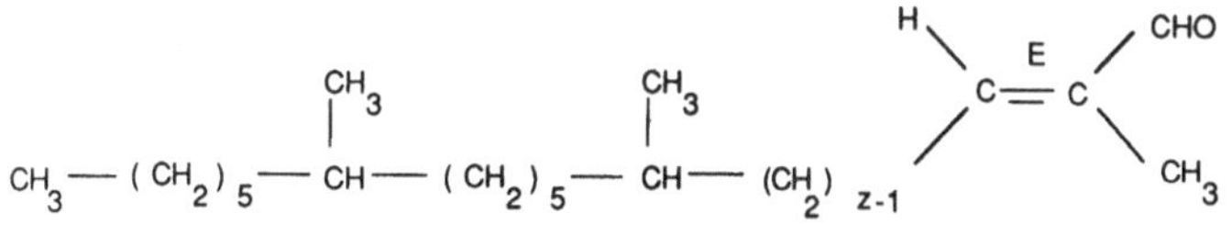

$$CH_3-(CH_2)_5-\underset{\underset{CH_3}{|}}{CH}-(CH_2)_5-\underset{\underset{CH_3}{|}}{CH}-(CH_2)_{z-1}-\underset{H}{\overset{}{\diagdown}}C\overset{E}{=}C\underset{CH_3}{\overset{CHO}{\diagup}} \quad \textbf{(77)}$$

$$CH_3-(CH_2)_5-\underset{\underset{CH_3}{|}}{CH}-(CH_2)_5-\underset{\underset{CH_3}{|}}{CH}-(CH_2)_z-\underset{\underset{CH_3}{|}}{CH}-CO_2H \quad \textbf{(78)}$$

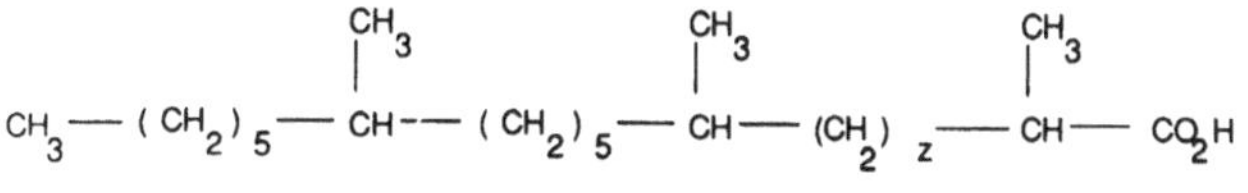

z : odd, 5 to 11

usual unbranched even acids exhibited, when analyzed as their methyl esters, a base peak at m/z 88 indicative of α-methyl substitution.

The location of the two "in-chain" methyl substitutions in series (76), (77) and (78) was deduced from the GC-EIMS of the common derived alkanes obtained by reduction to alcohols, followed by reductive removal of the hydroxyl groups by HI/P and subsequent treatment with Zn/HCl.

Structural considerations suggest a close biosynthetic relationship between these non-isoprenoid methyl-branched metabolites. Two possible origins can be considered for the methyl groups: (i) from methionine as in the monomethylalkanes synthesized by blue-green algae (63); (ii) from insertion of methylmalonate instead of malonate during the elongation process, as in the multiple-branched mycocerosic acids produced by mycobacteria (64).

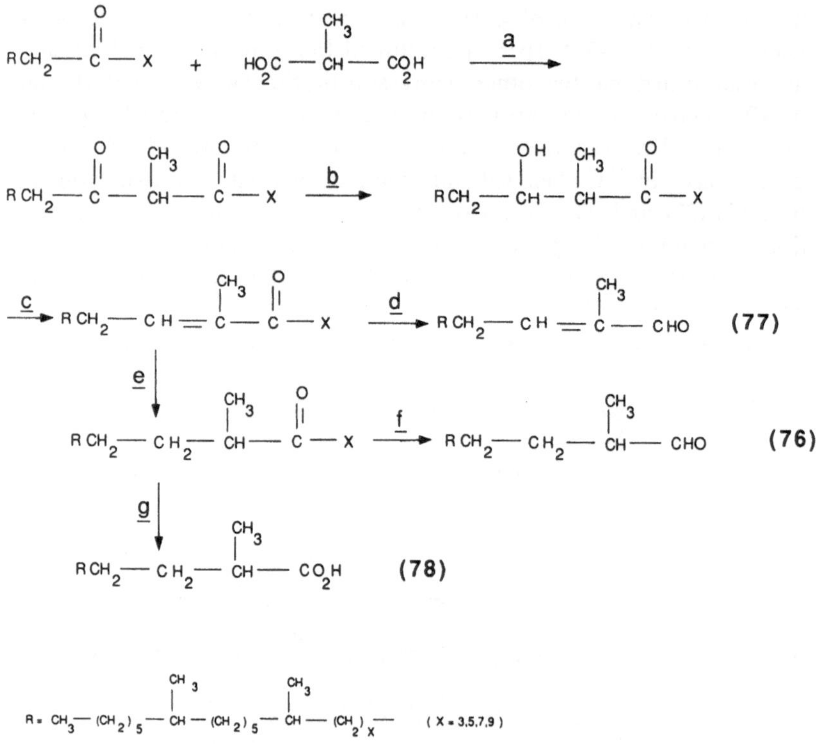

$$RCH_2-\overset{\overset{O}{\|}}{C}-X \;+\; HO_2C-\overset{\overset{CH_3}{|}}{CH}-CO_2H \quad\xrightarrow{\;\underline{a}\;}$$

$$RCH_2-\overset{\overset{O}{\|}}{C}-\overset{\overset{CH_3}{|}}{CH}-\overset{\overset{O}{\|}}{C}-X \quad\xrightarrow{\;\underline{b}\;}\quad RCH_2-\overset{\overset{OH}{|}}{CH}-\overset{\overset{CH_3}{|}}{CH}-\overset{\overset{O}{\|}}{C}-X$$

$$\xrightarrow{\;\underline{c}\;}\quad RCH_2-CH=\overset{\overset{CH_3}{|}}{C}-\overset{\overset{O}{\|}}{C}-X \quad\xrightarrow{\;\underline{d}\;}\quad RCH_2-CH=\overset{\overset{CH_3}{|}}{C}-CHO \quad (77)$$

$$\Big\downarrow \underline{e}$$

$$RCH_2-CH_2-\overset{\overset{CH_3}{|}}{CH}-\overset{\overset{O}{\|}}{C}-X \quad\xrightarrow{\;\underline{f}\;}\quad RCH_2-CH_2-\overset{\overset{CH_3}{|}}{CH}-CHO \quad (76)$$

$$\Big\downarrow \underline{g}$$

$$RCH_2-CH_2-\overset{\overset{CH_3}{|}}{CH}-CO_2H \quad (78)$$

$$R=CH_3-(CH_2)_5-\overset{\overset{CH_3}{|}}{CH}-(CH_2)_5-\overset{\overset{CH_3}{|}}{CH}-(CH_2)_x- \quad (X=3,5,7,9)$$

Fig. 17. Proposed biosynthetic relationship between the methyl-branched metabolites (76), (77) and (78). a: CO₂ elimination; b: β-keto acyl reduction; c: dehydration; d: acyl reduction; e: α,β reduction; f: acyl reduction; g: free fatty acid release

The location of the methyl groups on odd-numbered carbons of the hydrocarbon chains relative to the terminal CH_3, and the existence of the α-unsaturated aldehydes (**77**) would be rather in favour of the latter assumption. Thus, as shown in Fig. 17, the α-unsaturated fatty acyl derivatives resulting from a third condensation with methyl-malonate might undergo reduction either of the acyl group to give (**77**), or of the unsaturation to form saturated acyl intermediates; the latter would in turn be reduced or released to yield (**76**) and (**78**), respectively. The fact that all the above analyses were performed on cultures free from bacteria or fungi ascertained that (**76**)–(**78**) are true metabolites of *B. braunii* although they are quite unusual for green microalgae.

4. Macromolecular Lipids

4.1 B. braunii Rubber

The occurrence of a rubbery material in *B. braunii* was recently recognized in the strain from the Austin collection (A race) (*61*). This rubber is markedly different from the insoluble abiotic polymer obtained after prolonged exposure to air of botryococcene-containing *Botryococcus*; indeed, it was isolated from freshly dried biomass by extraction with chloroform. The precipitate obtained by addition of methanol or acetone to the chloroform extract (Fig. 1) was purified by repeated solution in $CHCl_3$ and subsequent addition of polar solvent until a white material was obtained. The polymer analyzed by size exclusion HPLC on an ultraStyragel column using THF as eluent showed a broad peak corresponding to a mass distribution ranging from 3.3×10^4 to 1.3×10^6 daltons, with a maximum of *ca* 1.6×10^5 daltons.

The IR, ^{1}H- and ^{13}C-NMR spectra revealed signals for an α-unsaturated, α-branched aldehyde, as previously found in the botryals, with *E* configuration for the conjugated carbon–carbon double bond. The NMR spectra also showed resonances for two *cis* non-conjugated double bonds in a polymethylene chain; no signals characteristic of aliphatic chain ends were observed. On ozonolysis, the polymer yielded three n-C_8, -C_9 and -C_{14} diacids in almost equal amounts and trace amounts of n-C_9 and -C_{16} monoacids, corresponding likely to the cleavage of the double bonds located in chain ends. The combination of these data with those from elemental analysis suggested the occurrence of a C_{32} monomer unit according to structure (**79**).

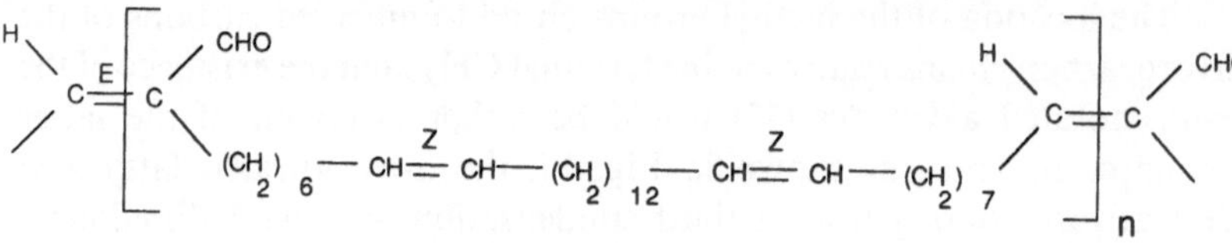

$$H_2C=\overset{E}{C}-\overset{CHO}{C}\left[(CH_2)_6-CH\overset{Z}{=}CH-(CH_2)_{12}-CH\overset{Z}{=}CH-(CH_2)_7\right] \quad H_2C=C-\overset{CHO}{C}$$

(79)

Preliminary feeding experiments performed with [1-^{14}C] and [10-^{14}C] oleic acid followed by radio-GC analysis of the ^{14}C labelled acids obtained on ozonolysis showed that oleic acid is a precursor of the rubber. Thus, after ozonolysis of the two ^{14}C-labelled rubbers radioactivity was found associated, in each case, with the n-C_{14} diacid fragment,

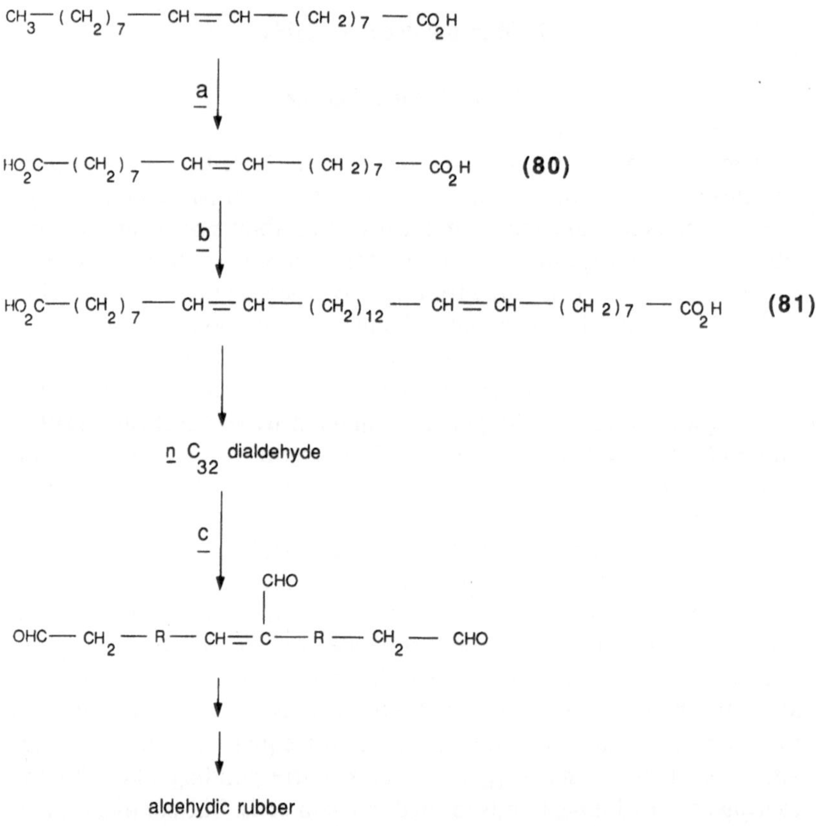

$$CH_3-(CH_2)_7-CH=CH-(CH_2)_7-CO_2H$$

$$\downarrow a$$

$$HO_2C-(CH_2)_7-CH=CH-(CH_2)_7-CO_2H \quad \textbf{(80)}$$

$$\downarrow b$$

$$HO_2C-(CH_2)_7-CH=CH-(CH_2)_{12}-CH=CH-(CH_2)_7-CO_2H \quad \textbf{(81)}$$

$$\downarrow$$

$$\underline{n}\ C_{32}\ \text{dialdehyde}$$

$$\downarrow c$$

$$OHC-CH_2-R-CH=\overset{CHO}{C}-R-CH_2-CHO$$

$$\downarrow$$

$$\downarrow$$

aldehydic rubber

Fig. 18. Proposed biosynthetic pathway of *B. braunii* rubber. *a*: ω-oxidation; *b*: elongation/dehydrogenation; *c*: aldolisation/dehydration

$$R=-(CH_2)_6-CH=CH-(CH_2)_{12}-CH=CH-(CH_2)_7-$$

suggesting that ω-oxidation of oleic acid yielding the symmetrical n-C_{18} diacid (**80**) had occurred before elongation into very long chain diunsaturated fatty diacids (**81**) (Fig. 18). Then, the derived dialdehydes would have undergone successive condensations followed by dehydration of the resulting aldols. Polymerization would have ended by an aldol reaction with a monoaldehyde. So far, the aldehydic rubber isolated from the A race of *B. braunii* has no equivalent in the plant kingdom. Preliminary investigations indicate the presence of rubbers of still undetermined structures in the B and L races.

4.2 Resistant Biopolymers from Outer Walls (PRB)

B. braunii outer walls are able to survive the successive degradations generally used for the isolation of sporopollenins which constitute a class of biopolymers occurring in the exine of spores and pollens. These latter biopolymers which would originate from the oxidative polymerization of carotenoids and/or carotenoid esters are insoluble in organic solvents and resistant to non-oxidative chemical attacks.

After hexane extraction and potassium hydroxide and then phosphoric acid treatments (Fig. 1), *B. braunii* outer walls exhibit, on examination in an electron microscope, an unchanged organization (Plate 2). The complete degradation of the outer walls by the action of chromic acid, established that they do not contain any silica which is known to be sometimes associated with a resistant organic polymer in algal walls.

A chemically resistant biopolymer termed PRB A was first discovered in the A race (*66*). It is important to specify that all studies reported to date on PRB A have been performed on algal residues still containing the rubber. Elemental composition of PRB A indicated a markedly lower oxygen content than observed in classical sporopollenins (atomic O/C ratio of *ca* 0.09 instead of *ca* 0.31 in *Lycopodium clavatum* sporopollenin) and a slightly higher hydrogen content (atomic H/C ratio of *ca* 1.73 instead of *ca* 1.53). The gross structural features of PRB A were identified using IR and high resolution solid state ^{13}C-NMR spectroscopy. The IR spectrum indicated a lower level of hydroxyl and methyl groups when compared with sporopollenins; the presence of some ether bridges was supported by two bands at 450 and 1160 cm^{-1}. Moreover, a substantial absorption at 720 cm^{-1} which may originate both from disubstituted *cis* double bonds and from long methylenic chains sharply distinguished PRB A from sporopollenins. A complex absorption between 1800 and 1550 cm^{-1} resulted from C=C double bonds and various carbonyl functions. The above observations were confirmed by high resolution

50   P. Metzger, C. Largeau, and E. Casadevall

solid state ^{13}C-NMR spectrometry which showed that PRB A contains mainly saturated carbons without heterosubstituents, giving rise to a prominent broad band from δ 20 to 40; the maximum at δ 29 was ascribable to long methylenic chains. Minor peaks in the spectrum were attributed to alcohol and ether functions at δ 60–75; C=C bonds at δ 110–130 and ester functions at δ 165–180.

Important structural information was also obtained from pyrolysis of PRB A at 400 °C carried out under an helium flow so as to minimize secondary reactions (67). GC-EIMS of the pyrolysate fractions obtained by CC and $AgNO_3$-SiO_2 TLC showed the predominance of hydrocarbons (ca 55% of the total pyrolysate). Regular series of C_{13} to C_{31} n-alkanes and n-alk-1-enes, formed by cracking of C–C bonds, are the major constituents of this fraction. They are accompanied by minor series of n-alkylbenzenes, n-alkyl- and n-alkenylcyclohexanes and n-trans-alkenes. Pyrolysis also provided a complex mixture of unidentified ketones and a series of n-fatty acids dominated by palmitic and oleic acids. The recovery of fatty acids on pyrolysis of PRB A, although isolation of this resistant material required drastic basic and acid treatments, indicates that the corresponding esters are sterically protected in the polymeric network.

Identification of these pyrolysis products added to spectroscopic information indicated that PRB A is built up mainly from long, unbranched, hydrocarbon chains, probably linked by ether bridges and containing some double bonds along with hydroxyl groups and ester functions. Accordingly, the resistant polymer from the A race of B. braunii cannot derive from carotenoids and in this respect cannot be regarded as a sporopollenin. This lack of a relationship to the sporopollenins was further confirmed by feeding experiments with [10-^{14}C] and [9,10-^{3}H] oleic acid, the radioactivity thus incorporated being chiefly located in the hydrocarbon fraction obtained by pyrolysis of labelled PRB A (68, 69). This result established the involvement of oleic acid in the formation of the long hydrocarbon chains building up the polymeric network.

In spite of all these results, the precise origin and structure of PRB A are still to be established. However the ether lipids (72)–(75) which were identified in the external lipids of the A race are very likely implicated as important building blocks in the formation of PRB A. Indeed, the structural features of these ether lipids (very long hydrocarbon chains, ether bridges and hydroxyl and ester groups) are consistent with those of PRB A and, as indicated in part 3.4, a polymeric network could be easily generated via cross-linking of such compounds. Moreover, a clear absorption at 1690 cm^{-1} in the IR spectrum of PRB A probably reflects the occurrence in the polymeric network of some botryal moieties such as the one of (75). Recently, *n*-alkylphenols, (chain length from C_4 to C_{12}), were identified in the pyrolysate of PRB A (*70*); although they were minor compounds, their detection could reflect the implication of some phenolic ether lipids such as (74) in the formation of the resistant polymer.

The question arises finally whether or not the aldehydic rubber also contributes to the formation of PRB A. Indeed, when the algal biomass is not extracted with $CHCl_3$, all the rubber remains definitely included in the final residue and the PRB A content (*ca* 10% of dry biomass) is twice that found when rubber is removed by $CHCl_3$ extraction prior to basic and acid hydrolysis (*65*). However, this observation does not entirely rule out the possible incorporation of some rubber, perhaps in reticulated form, into PRB A.

From elemental analysis, IR and ^{13}C-NMR spectroscopy and GC-MS investigations of pyrolysates, PRB B isolated from the B race (*ca* 10% of dry biomass), exhibits very similar structural features when compared to PRB A; only slightly more important methyl-branching is observed in the former (*71*). In addition, feeding experiments with radio-labelled compounds confirmed the involvement of long unbranched, hydrocarbon chains in the biopolymer network. Indeed, sodium DL [2-^{14}C] mevalonate, a botryococcene precursor, was poorly incorporated, while [10-^{14}C] oleic acid resulted in strongly labelled PRB B (*69*, *72*); in this latter case radioactivity was recovered both in the hydrocarbons and the fatty acids released on pyrolysis. To date the origin of PRB B is still obscure, no ether lipids similar to those found in the A race have been so far isolated from the B race.

The resistant biopolymer PRB L occurring in the L race (*ca* 30% of dry biomass), shows a markedly different chemical structure when compared with PRB A and PRB B (*73*). IR and ^{13}C-NMR spectra pointed to an important contribution of methyl groups relative to methylenes in the former resistant material; ether and ester functions, hydroxyl groups and unsaturations were also detected. The high level of methyl-branching occurring in PRB L was confirmed by GC-MS identification of the pyrolysis products. Numerous isoprenoid compounds were thus identified in the hydrocarbon subfraction (63% of the total pyrolysate) and in the ketone subfraction (22% of the total pyrolysate) as well.

The hydrocarbon subfraction is dominated by C_{40} isoprenoid compounds (*ca* 40% of total hydrocarbons); it also comprises several homologous series of alkanes and alkenes up to C_{30}. In these series, regular head-to-tail C_{13}–C_{21} isoprenoid alkanes and C_9–C_{24} isoprenoid alkenes occur in substantial amounts, whereas C_{12}–C_{30} *n*-alkanes are only minor components. The C_{40} isoprenoid compounds were identified and their structure established on the basis of their fragment ions observed in EIMS (*74*). All these compounds exhibit a central or a terminal benzene ring such as in (**82**) and in (**83**), the two major hydrocarbons of the C_{40} group.

(**82**)

(**83**)

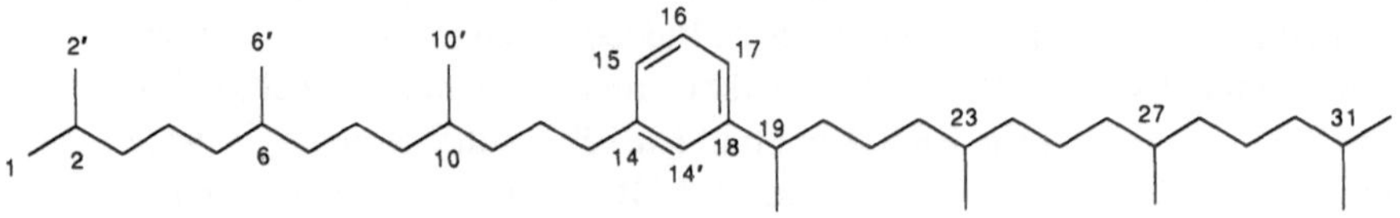

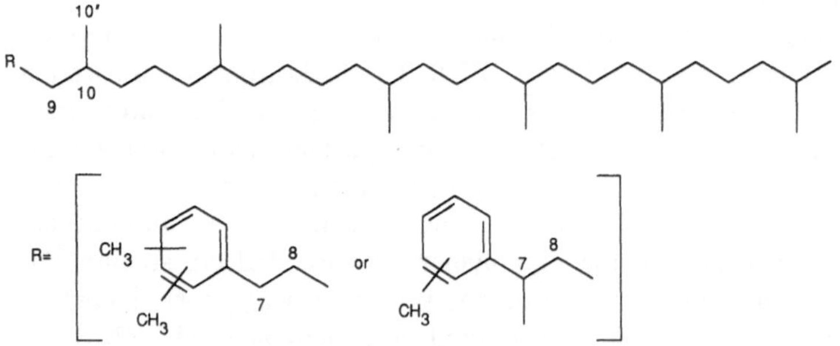

The ketone subfraction is dominated markedly by a monounsaturated C_{40} isoprenoid ketone (**84**), the structure of which is not still definitely established with respect to the location of the unsaturation; three diunsaturated C_{40} ketones (**85**)–(**87**) were also identified. In addition this subfraction comprises C_{11}–C_{22} isoprenoid ketones, the most important of which is the C_{18} compound (**88**).

(**84**)

(**85**)

(**86**)

References, pp. 63–70

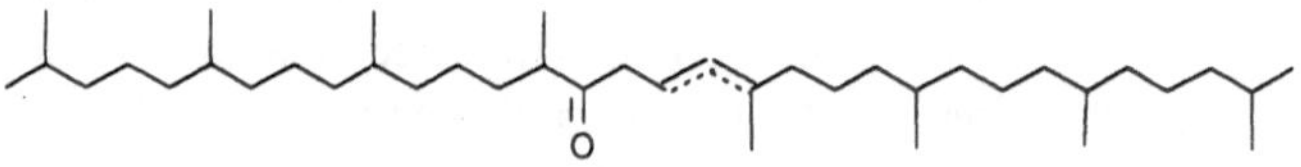

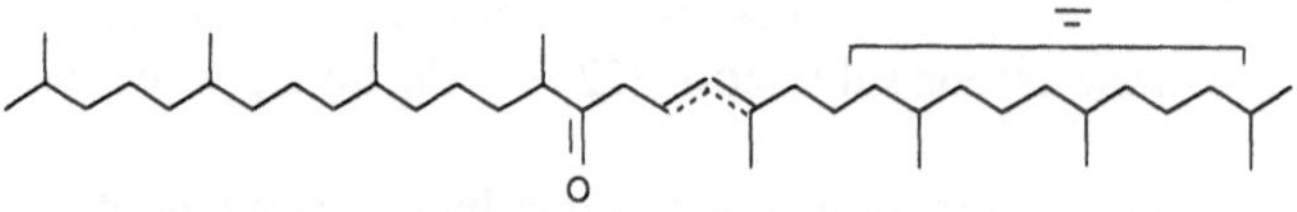

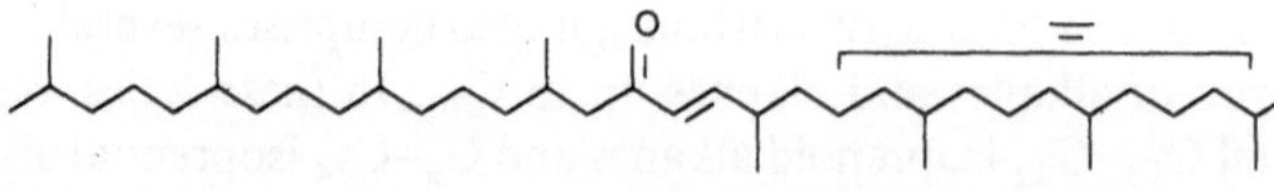

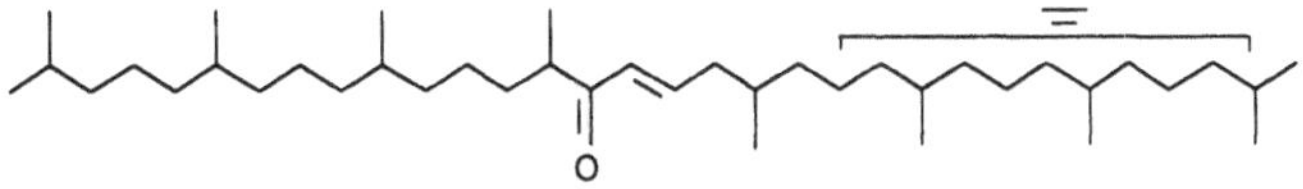

(87)

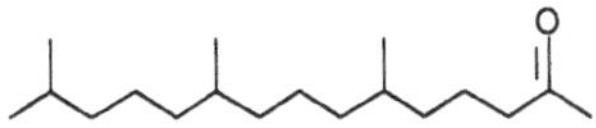

(88)

The carboxylic acid subfraction is essentially composed of n-C_{12} to C_{18} even carbon numbered fatty acids, with palmitic acid as the major constituent (75). Isoprenoid acids (C_{14}, C_{15}, C_{17} and C_{19}) were also detected in this subfraction; the most abundant is the C_{15} acid (89). Such acids, whose structure was established by means of the fragmentation patterns in EIMS are rarely found in nature.

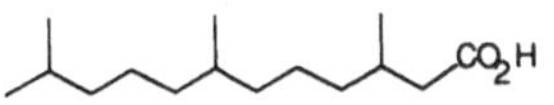

(89)

On the basis of their carbon number and chemical structure, all the very long chain compounds identified in the 400 °C pyrolysate of PRB L can be related to lycopadiene (45) or to lycopane, its hydrogenated derivative (90). Thus, structure (82) is consistent with cyclization occurring in a lycopane skeleton between C(14′) and C(18). In the case of (83), the aromatic ring would originate from cyclization between C(2) and C(6′) or C(2′) and C(6) with subsequent aromatization and migration of a methyl group. All the very long chain ketones exhibit the central tail-to-tail linkage between two phytyl-type units typical of the lycopane skeleton. Accordingly such ketones are directly related to lycopadiene (45), and their carbonyl function, located at C(15) or at C(16), probably originates from the thermal cleavage of C–O–C cross-links. No direct relationship exists between lycopadiene on one hand and ketone (88) or isoprenoid acid (89) on the other hand, but such regular isoprenoid structures could arise from cleavage of a lycopane skeleton. The observation of large amounts of C_{40} "lycopane-type", isoprenoid products, demonstrates clearly that PRB L comprises mainly C_{40} units, likely linked by ether bridges.

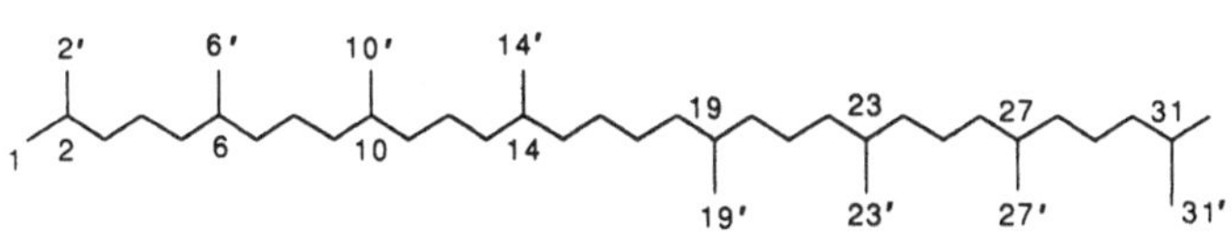

(90)

In the three races of *B. braunii*, the chemical structure of PRB is therefore based on a polymeric network of long hydrocarbon chains likely cross-linked by ether bridges. Furthermore in the A and L races close structural correlations exist between these chains and the hydrocarbons produced by *B. braunii*. By contrast a complete lack of relationship is noted in the case of the B race.

III. Geochemical Implications of *B. braunii* Composition

1. Role of PRB in Fossilization

Recent comparative studies on the chemical composition of extant *B. braunii* and of their fossil counterparts afforded important geochemical information, especially with regard to the origin and to the mode of formation of kerogens (the insoluble organic matter dispersed in sedimentary rocks). The thermal degradation of kerogens associated with progressive burial results in the formation of petroleum and natural gas (*76, 77*). In addition the morphological and chemical features of kerogens can provide a wealth of information on the nature and composition of the contributing organisms (chiefly microalgae in a number of cases) and on the environmental conditions prevailing at the ·time of their deposition. Accordingly a precise knowledge of the chemical structure and of the mode of formation of kerogens appears of a great importance.

Until the last few years it was considered that kerogens originated from the so-called "Degradation–Recondensation" pathway (*77*). According to this pathway, prolonged degradations, chiefly of bacterial origin, taking place in the upper layer of sediments would cause extensive alterations, including depolymerization, of biological macromolecules like proteins and polysaccharides. As a result of this initial degradative step the bulk of the deposited organic matter would be completely eliminated by mineralization into CO_2 and H_2O. However a minor part would escape mineralization owing to random polycondensation reactions. Such reactions would result in progressive insolubilization and increasing resistance to hydrolysis of the organic matter fixed in sediments, thus leading to the formation of an insoluble polycondensate (kerogen). According to the above pathway extensive alterations with respect to both the morphology and the chemical composition of the initial material should occur during fossilization. Furthermore, even under favourable conditions where the degradation processes are limited, only a low portion of the total input of organic matter to sediments, estimated to a few percent, would be transformed into kerogen. The occurrence of low levels of kerogen finely dispersed in a mineral matrix in a number of sedimentary rocks and the amorphous appearance of most kerogens when examined by light microscopy were therefore consistent with the above Degradation–Recondensation pathway.

However, some sedimentary rocks are known to contain very high levels of kerogen, nearly exclusively composed of massive accumulations of a single type of microfossil well retaining the morphological features of the initial organism. Torbanites are probably the most typical example of such organic-rich deposits. Furthermore, it has long been recognized that the fossil microorganisms building up the kerogen of torbanites exhibit a general organization similar to that of colonies in modern *B. braunii* (*78, 79*) (Plate 2). The chemical structure of torbanites was therefore examined by the spectroscopic and pyrolytic methods previously used for studying the resistant biopolymers (PRB) isolated from the outer walls of extant *B. braunii*. Very closely related structures were thus observed in torbanites and in the resistant biopolymers, PRB A and PRB B respectively, from the A and B races as illustrated by the solid state ^{13}C-NMR spectra in Fig. 19 (*36, 67, 68, 80–85*). This chemical evidence definitely confirmed the correlation between *B. braunii* and torbanites so far based only on morphological similarities. Moreover such observations revealed that the mode of formation of kerogen in torbanites was entirely different from the classical Degradation–Recondensation pathway. Owing to their very high resistance to bacterial degradation and to hydrolysis, PRB A and PRB B remained virtually unaffected, both quantitatively and qualitatively, during fossilization. The resistant material of the outer walls was thus selectively preserved and provided a major contribution to torbanites while the non-resistant constituents of *B. braunii* only afforded a very low contribution *via* the Degradation–Recondensation pathway. The involvement of this "Selective Preservation" process accounts for (i) the retention of *B. braunii* morphology on fossilization (due to the presence of PRB in outer walls, since the latter build up the matrix of colonies and are responsible for their typical organization), (ii) the remarkable efficiency of *B. braunii* for producing organic-rich deposits (the resistant wall material, *i.e.* a substantial fraction of the initial biomass, is entirely retained as kerogen), (iii) the very high oil potential of torbanites (the polymeric structure of PRB A and B is based on a network of long hydrocarbon chains, hence an abundant production of petroleum on thermal cracking at depth).

Subsequent studies revealed similar morphological and chemical correlations between various kerogens and the resistant biopolymers

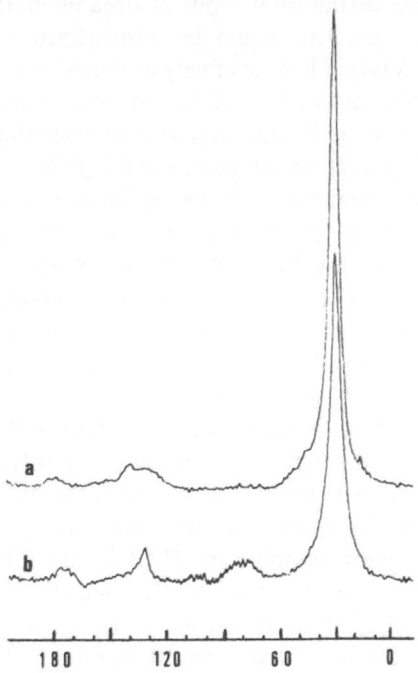

180   120   60   0

Fig. 19. Solid state ^{13}C NMR spectra of a torbanite (a) and of PRB from the B race (b), by courtesy of Dr. S. Derenne (Paris)

building up cell walls or protective layers in several groups of organisms including microalgae, cyanobacteria and higher plants (86–93). The Selective Preservation pathway was thus shown to play a widespread and often prominent role in kerogen formation in a number of petroleum source rocks. On the basis of such studies, a general concept of kerogen formation *via* the selective preservation of resistant macromolecules was recently worked out (94, 95).

2. Role of *B. braunii* Hydrocarbons as Biomarkers

Analysis of oils or of rock extracts reveals the presence of products, mostly hydrocarbons, the basic structure of which is related to that of typical lipoid constituents of living organisms. During fossilization the latter constituents undergo substantial alteration, including the loss of functional groups and the reduction of olefinic unsaturations, but their carbon skeleton is retained or affected only to a minor extent. While such "biomarkers", also often termed geochemical fossils, usually occur in very low amounts when individual compounds are considered, they have been extensively studied since they provide information on the nature of the initial organisms and on their deposition conditions (*96*). Biomarkers usually show poor specificity and are generally related to an entire group of organisms, or even to several distinct groups. Thus very long chain linear compounds, with odd numbers of carbon atoms in the range C_{25}–C_{33}, are major constituents of higher plant waxes (*97*). The presence of very long chain *n*-alkanes with an odd number of carbon atoms predominating is therefore generally considered as indicative of terrestrial origin (initial organic matter mainly derived from land plants) of the sample under consideration. However fossilization of the A race of *B. braunii* should also result, by reduction of *n*-alkadienes and trienes, in the formation of very long chain *n*-alkanes showing the same predominance of odd numbered carbon chains. Accordingly this group of biomarkers cannot alone afford conclusive evidence for the involvement of either higher plants or of the *B. braunii* A race in the formation of fossil materials. The hydrocarbons derived from the non-classical lipids of the A race of *B. braunii* (like **57** and **58**) should, on the contrary, provide specific biomarkers. Such hydrocarbons have so far not been identified in oils or rock extracts. However this could merely reflect the experimental problems (recovery, appropriate GC conditions) associated with the analysis of high molecular weight compounds; in fact, in nearly all previous studies about biomarkers only hydrocarbons with carbon numbers up to C_{40} have been examined.

Botryococcenes have been detected exclusively in the B race of *B. braunii*; the derived alkanes, the botryococcanes, can therefore be considered as specific markers of *B. braunii*. This is the only one-to-one correspondence so far established between a series of biomarkers and a living organism. C_{34} botryococcane, (**91**), was first discovered in two Sumatran oils in a concentration (up to 1.4% of total crude oil) markedly higher than ever observed before for any individual biomarker (*96, 98*). This alkane showed the same carbon skeleton as the C_{34} botryococcene (**16**) occurring in large amount in several strains of the B race of *B. braunii*. The presence of botryococcane in the above oils revealed a substantial contribution of the B race of *B. braunii* to the formation of the source rocks from which these oils were derived. C_{34} botryococcane was also identified, in high concentrations, in Australian coastal bitumens

(91)

(99). C_{31} and C_{33} botryococcanes dominated the aliphatic hydrocarbon fraction isolated from the extract of the Maoming oil shale of China *(100)*. The abundance of botryococcanes observed in all these studies is consistent with the occurrence of very high amount of botryococcenes in the B race of *B. braunii*. Nevertheless, in spite of extensive studies aimed at identifying these species-specific biomarkers, botryococcanes have been detected in very few geological samples *(81)* while *Botryococcus* remains are widespread in a number of kerogens. This limited occurrence could reflect either a quite poor resistance of the botryococcenes to the pronounced bacterial and chemical degradations taking place on a geological time scale (*i.e.* for thousands of years) during sedimentation under oxic conditions as indicated by studies on the first stages of *B. braunii* fossilization *(81)*, or a large contribution of the A race to numerous *Botryococcus* deposits.

As previously discussed, the L race of *B. braunii* produces substantial amounts of a single hydrocarbon, lycopadiene (**45**). As a result, lycopane (**90**) should be a typical biomarker for this race. However, unlike the botryococcenes, this is not a species-specific biomarker since lycopane can also be derived by reduction of acyclic C_{40} carotenoids which are relatively common in various microorganisms and plants. The presence of lycopane was reported, along with C_{31} and C_{33} botryococcanes, in the Maoming oil shale *(100)*. Furthermore, observations on algal populations in some lakes point to the quite common co-occurrence of the B and L races of *B. braunii* (Table 1). The presence of both lycopane and botryococcanes should therefore reflect the contribution of the L race, in addition to the B race, to the formation of the Maoming oil shale.

IV. *B. braunii* Biotechnology

As indicated by the occurrence of conspicuous blooms, *B. braunii* can exhibit a profuse growth in nature. Some of the parameters which may initiate such a growth were examined by WAKE and HILLEN *(2)* by their study of a large and persistent bloom in Darwin Reservoir, Australia. Nevertheless the combination of environmental conditions leading to *B. braunii* blooms is still far from being understood. As regards laboratory cultures, the only available information on the nutrient requirements of this species and the suitable range of culture parameters was provided, until recently, by the pioneering work of CHU *(101)*; the first cultures resulted in slow growth with a mean doubling time of *ca* one week *(21, 102)*. Studies on endogenous respiration and on photosynthetic activity *(102, 103)* indicated that the rate of the general metabolism of *B. braunii* is similar to that of fast growing algae (mean doubling time of 6–8 h) like *Chlorella*. The slower growth of the former species was thus considered as related, at least in part, to the large scale production of hydrocarbons not available as a food reserve. However a substantial improvement in growth rate was achieved under "air lift" conditions (mean doubling time of *ca* 2 days) with the three different races of *B. braunii* (*11, 12, 20, 103–105*). Heterotrophic growth of *B. braunii* with a suitable carbon source like mannose also reduced the mean doubling time to *ca* 3 days (*106*). Labelling experiments with $^{14}CO_2$ recently suggested (*107*) that *B. braunii* assimilates inorganic carbon by the C-3 pathway.

1. Relationship Between Hydrocarbon Production and Physiological State

Large accumulations of lipids, especially triacylglycerols, are commonly observed in various algae when cell division is blocked or strongly inhibited while carbon fixation still goes on, *i.e.* during the stationary phase of batch cultures (*108–110*). Very high levels of total lipids, up to 85% of biomass, were thus obtained (*108*) in some N- or P-deficient green microalgae. The relationships between hydrocarbon formation and physiological state were therefore examined in the three races of *B. braunii*. During the exponential growth stage of batch cultures, hydrocarbon production was shown to be faster than production of the other cell constituents. Accordingly the hydrocarbon level relative to total biomass substantially increases during this stage (*e.g.* up to 44% in a strain of the A race from a relatively low value of 16% in the inoculum) (*104*). Afterwards, during the deceleration and stationary phases, a reverse trend occurs and the level of hydrocarbons is lowered. In batch cultures of the three races the maximal hydrocarbon productivity is thus obtained from the active growth phases (*11, 20, 103, 104, 111*). Furthermore no accumulations of hydrocarbons occurred in N- and P-deficient cultures of the B race (*12*). Continuous cultures of the A race confirmed that the large hydrocarbon production typical of *B. braunii* does not require inhibition of cell division but, on the contrary, actively dividing cells. In addition, comparison of "standard" and "air lift" batch cultures indicated that the large improvement in growth rate obtained under the latter conditions is also associated with a substantial increase in the hydrocarbon level of the produced biomass, hence a very sharp rise in hydrocarbon productivity (*103*). On the basis of all the above observations, hydrocarbon accumulation by *B. braunii* and triacylglycerol accumulation by various algae appear therefore to be of an entirely different nature, the former being associated with active growth conditions.

The influence of pronounced changes in growth conditions and in physiological stage on the nature and the relative abundance of *B. braunii* hydrocarbons was also examined. With the A race, no significant variations were detected (*111*); in sharp contrast the composition of the hydrocarbon fraction, botryococcenes, produced by the B race is highly sensitive to growth conditions (*11, 12*).

2. Influence of Various Culture Parameters on Hydrocarbon Production

2.1 Medium Composition

Nitrate was shown to be limiting in the "air lift" cultures carried out with a modified CHU 13 medium (*111*). Progressive increase in nitrate initial concentration thus results in a more extended exponential phase, until another nutrient becomes in turn limiting. This adjustment of the nitrate supply allows for a substantial increase in final biomass and hydrocarbon production in batch cultures, *ca* 75 and 60%, respectively.

Since phosphate is not limiting, higher initial concentrations do not cause any changes in the growth curve or in the final biomass from batch cultures of *B. braunii* (*104*). However the lipid content of algae is known to be influenced by the P/N ratio in the growth medium and the hydrocarbon production of *B. braunii* is somewhat improved (*ca* 10%) following a twofold increase in phosphate supply (*104*).

No adverse effect of salinity up to that of seawater was observed on *B. braunii* growth and addition of 6% NaCl in the medium resulted in a substantial increase of the level of total lipids (*ca* 50% of biomass instead of *ca* 35%) (*112*).

2.2 Light

Adjustments in light intensity cause marked increases in both total biomass and hydrocarbons. A more than twofold increase in hydrocarbon production was thus achieved in the A race, through optimization of light intensity (*113*).

2.3 B. braunii Strain

A number of strains of the A and B races were isolated from samples collected in various countries (Table 1). When laboratory-grown under similar conditions the strains of a given race showed marked differences in both hydrocarbon and biomass production and in the level of hydrocarbons relative to total biomass (*11, 105*). Thus, on comparison of A race strains a thirty-fold difference in hydrocarbon level was noted between low and high productivity strains (*14*).

2.4 Associated Bacteria

Various bacteria were combined with an axenic strain of the A race and their influence on *B. braunii* growth and hydrocarbons was examined (*114–116*). Most of the tested microorganisms exert a strong detrimental influence; however *Flavobacterium aquatile* exerts large promoting effects, both on the total algal biomass and on hydrocarbon production, probably as a result of the release of stimulating substances in the culture medium.

3. Immobilized Cultures

The influence of immobilization was examined following direct entrapment of whole cells of *B. braunii* in alginate gel and polyurethane foams; adsorption on preformed foams was also employed (*117–121*). All these experiments were carried out using batch cultures of the A race.

The growth rate of the cells embedded in alginate beads is somewhat lower when compared with free cells (mean doubling time of 2.5 days instead of 1.7 day). Electron microscopy indicated that such a decrease originates from steric constraints since the colonies, observed just after entrapment, appeared located in small size pores. This growth limitation results in some diversion of *B. braunii* metabolic activity towards secondary compounds and the production of hydrocarbons is improved by *ca* 20% relative to free controls. Immobilization in alginate was also shown to exert a protective influence on the photosynthetic system of *B. braunii*, especially against the early degeneration occurring under "air lift" conditions in free cultures. Such a degeneration causes a sharp decrease in chlorophyll content and a gradual disorganization of cells and becomes rapidly irreversible if stationary cells are not inoculated into a fresh medium. The onset of this process is also associated with a sharp decrease in hydrocarbon production (*104*). When *B. braunii* is immobilized in alginate, early degeneration is markedly delayed and slowed down relative to free cells. This type of immobilization therefore exerts both a stimulating influence on hydrocarbon production and a protective influence on the general metabolic activity of *B. braunii*.

Direct entrapment of *B. braunii* in various polyurethane foams revealed high toxicity of most of the tested polymers and a complete loss of viability occurred in numerous cases. However, a few polymers appear to be less toxic and some cells can survive entrapment; nevertheless these showed poor metabolic activity and a *ca* fivefold decrease in hydrocarbon production was noted relative to free cells. In sharp contrast, no toxicity was detected following immobilization through adsorption on preformed foam pieces. Furthermore the total biomass and hydrocarbon productivities then obtained from the immobilized cultures are similar to the high values typical of controls.

4. Hydrocarbon Recovery

As previously noted, the bulk (*ca* 95%) of *B. braunii* hydrocarbons is stored in the thick outer walls surrounding the basal part of cells. A substantial proportion of these external hydrocarbons can be recovered by mild mechanical treatments and the cells used again for a new run (*103*). However this method provides a water-rich emulsion and furthermore is not suitable for processing immobilized cultures. The efficiency of recovery extraction with biocompatible solvents was therefore examined. To this end 18 candidate solvents were selected from a database of about 1500 solvents and tested (*122*). A strong negative relationship between solvent polarity and biocompatibility was thus observed. Nine of the candidate solvents proved to be biocompatible with *B. braunii* and provided high recovery yields (up to 70% with *n*-hexane) when contacted with a wet cell paste obtained by filtration. In addition, no detrimental effects on growth and hydrocarbon production were detected on subsequent cultures of the contacted algae when compared to untreated controls (*123*). A similar lack of impairment of *B. braunii* biological activity was observed in immobilized cultures. Moreover it was noted that immobilized cells, particularly those adsorbed on polyurethane foams, afforded higher recovery yields than filtered algae, especially when short contact times are considered (*123*).

It is now well documented that immobilized cultures offer numerous advantages relative to free cultures including a pronounced protective influence. In addition, the studies concerned with *B. braunii* revealed that immobilization can stimulate hydrocarbon production and increase the efficiency of recovery with biocompatible solvents.

V. Conclusion

The study of a large number of strains of *B. braunii* has revealed that unusually high amounts of lipids, especially hydrocarbons, are produced by this ubiquitous microalga. A considerable diversity in the chemical structure of such compounds has also been observed and several groups of "non-classical" lipids, so far not detected in other organisms, like botryals and ether lipids, have been identified. Specific macromolecular lipids, including insoluble and chemically resistant biopolymers building up outer walls, PRB, have also been shown to occur in *B. braunii*. The above mentioned non-classical lipids are likely involved, as basic units, in the formation of PRB. Indeed an extended array of compounds of increasing molecular weight and decreasing solubility, ranging from simple ether lipids to PRB, seems to be present in *B. braunii*.

Fossil colonies of *Botryococcus* are commonly found in sedimentary rocks ranging in age from late Proterozoic (*ca* 10^9 years) to Recent. Furthermore a number of organic-rich deposits with a very high oil potential are composed mainly of accumulations of such microfossils. In fact, the formation of large amounts of PRB accounts for the geochemical importance of *B. braunii*. Owing to their high resistance to degradations, PRB are directly fossilized via selective preservation; *B. braunii* thus makes a large contribution to kerogen formation in various deposits including some petroleum source rocks. Moreover, the hydrocarbons produced by the different races of *B. braunii* are an important source of biomarkers, including botryococcanes, *i.e.* the only species-specific family of biomarkers so far identified.

Finally, *B. braunii* presently appears to be the best candidate species for the possible production of renewable hydrocarbons, by large scale culture of photosynthetic organisms. The non-isoprenoid and the isoprenoid hydrocarbons produced by the different races of *B. braunii* might thus be used as multipurpose feedstocks, either as such or after chemical transformations.

References

1. AARONSON, S., T. BERNER, K. GOLD, L. KUSHNER, N.J. PATNI, A. REPAK, and D. RUBIN: Some observations on the green planktonic alga, *Botryococcus braunii* and its bloom form. J. Plankton Res. **5**, 693 (1983).

2. WAKE, L.V., and W. HILLEN: Study of a bloom of the oil-rich alga *Botryococcus braunii* in the Darwin River Reservoir. Biotechnol. Bioeng. **22**, 1637 (1980).

3. BLACKBURN, K.B.: A reinvestigation of the alga *Botryococcus braunii* Kützing. Trans. Roy. Soc. Edinburgh **58**, 841 (1936).

4. Schnepf, E., and W. Koch: Über den Feinbau der Ölalge *Botryococcus braunii* Kützing (Chlorococcales). Bot. Jahrb. Syst. Pflanzengesch. Pflanzengeogr. **99**, 370 (1978).

5. Berkaloff, C., B. Rousseau, A. Coute, E. Casadevall, P. Metzger, and C. Chirac: Variability of cell wall structure and hydrocarbon type in different strains of *Botryococcus braunii*. J. Phycol. **20**, 377 (1984).

6. Bertrand, C.E., and B. Renault: *Pila bibractensis* et le boghead d'Autun. B. Soc. Hist. Nat. Autun. **5**, 160 (1892).

7. Zalessky, M.D.: Sur les nouvelles algues découvertes dans le sapropélogène du Lac Beloe et sur une algue sapropélogène. Rev. Gen. Bot. **38**, 31 (1926).

8. Swain, F.M., and J.M. Gilby: Ecology and taxonomy of Ostracoda and an alga from Lake Nicaragua. Publ. Stn. Zool. Napoli 33 (Suppl.), 361 (1964).

9. Maxwell, J.R., A.G. Douglas, G. Eglinton, and A. McCormick: The botryococ-cenes-Hydrocarbons of novel structure from the alga *Botryococcus braunii* Kützing. Phytochemistry **7**, 2157 (1968).

10. Brown, A.C., B.A. Knights, and E. Conway: Hydrocarbon content and its relation-ship to physiological state in the green alga *Botryococcus braunii*. Phytochemistry **8**, 543 (1969).

11. Metzger, P., C. Berkaloff, E. Casadevall, and A. Coute: Alkadiene- and botryococcene-producing races of wild strains of *Botryococcus braunii*. Phytochemis-try **24**, 2305 (1985).

12. Wolf, F.R., A.M. Nonomura, and J.A. Bassham. Growth and branched hydrocar-bon production in a strain of *Botryococcus braunii* (Chlorophyta). J. Phycol. **21**, 388 (1985).

13. Wake, L.V., and L.W. Hillen: Nature and hydrocarbon content of the alga *Botryo-coccus braunii* occurring in Australian freshwater lakes. Aust. J. Mar. Freshwater Res. **32**, 353 (1981).

14. Metzger, P., E. Villarreal-Rosales, E. Casadevall, and A. Coute: Hydrocar-bons, aldehydes and triacylglycerols in some strains of the A race of the green alga *Botryococcus braunii*. Phytochemistry **28**, 2349 (1989).

15. Knights, B.A., A.C. Brown, E. Conway, and B.S. Middleditch: Hydrocarbons from the green form of the freshwater alga *Botryococcus braunii*. Phytochemistry **9**, 1317 (1970).

16. Metzger, P., J. Templier, C. Largeau, and E. Casadevall: A *n*-alkatriene and some *n*-alkadienes from the A race of the green alga *Botryococcus braunii*. Phytochemistry **25**, 1869 (1986).

17. Metzger, P., E. Casadevall, and A. Coute: Botryococcene distribution in strains of the green alga *Botryococcus braunii*. Phytochemistry **27**, 1383 (1988).

18. Wolf, F.R., and E.R. Cox: Ultrastructure of active and resting colonies of *Botryococ-cus braunii* (Chlorophyceae). J. Phycol. **17**, 395 (1981).

19. Metzger, P., and E. Casadevall: Lycopadiene, a tetraterpenoid hydrocarbon from new strains of the green alga *Botryococcus braunii*. Tetrahedron Letters **28**, 3931 (1987).

20. Metzger, P., B. Allard, E. Casadevall, C. Berkaloff, and A. Coute: Structure and chemistry of a new chemical race of *Botryococcus braunii* (Chlorophyceae) that produces lycopadiene, a tetraterpenoid hydrocarbon. J. Phycol. **26**, 258 (1990).

21. Largeau, C., E. Casadevall, C. Berkaloff, and P. Dhamelincourt: Sites of accumulation and composition of hydrocarbons in *Botryococcus braunii*. Phytochem-istry **19**, 1043 (1980).

22. Largeau, C., E. Casadevall, and C. Berkaloff: The biosynthesis of long-chain hydrocarbons in the green alga *Botryococcus braunii*. Phytochemistry **19**, 1081 (1980).

23. METZGER, P., M. DAVID, and E. CASADEVALL: Biosynthesis of triterpenoid hydrocarbons in the B race of the green alga *Botryococcus braunii*. Sites of production and nature of the methylating agent. Phytochemistry **26**, 129 (1987).

24. VILLARREAL-ROSALES, E.: Unpublished results.

25. —: Thèse 3ème Cycle, Université de Technologie de Compiègne, 1990.

26. PAMBOU-TCHIVOUNDA, H. Unpublished results.

27. TEMPLIER, J., C. LARGEAU, and E. CASADEVALL: Mechanism of non-isoprenoid hydrocarbon biosynthesis in *Botryococcus braunii*. Phytochemistry **23**, 1017 (1984).

28. ———: Effect of various inhibitors on biosynthesis of non-isoprenoid hydrocarbons in *Botryococcus braunii*. Phytochemistry **26**, 377 (1987).

29. CHAN YONG, T.P., C. LARGEAU, and E. CASADEVALL: Biosynthesis of non-isoprenoid hydrocarbons by the microalga *Botryococcus braunii*. Evidence for an elongation–decarboxylation mechanism. Activation of decarboxylation. Nouv. J. de Chimie **10**, 701 (1986).

30. TEMPLIER, J., C. LARGEAU, and E. CASADEVALL: Non-specific elongation-decarboxylation in biosynthesis of *cis*- and *trans*-alkadienes by *Botryococcus braunii*. Phytochemistry **30**, 175 (1991).

31. COX, R.E., A.L. BURLINGAME, D.M. WILSON, G. EGLINTON, and J.R. MAXWELL: Botryococcene – a tetramethylated acyclic triterpenoid of algal origin. J. Chem. Soc. Chem. Comm. 284 (1973).

32. METZGER, P., E. CASADEVALL, M-J. POUET, and Y. POUET: Structures of some botryococcenes: branched hydrocarbons from the B race of the green alga *Botryococcus braunii*. Phytochemistry **24**, 2995 (1985).

33. HUANG, Z., C.D. POULTER, F.R. WOLF, T.C. SOMERS, and J.D. WHITE: Braunicene, a novel cyclic C_{32} isoprenoid from *Botryococcus braunii*. J. Am. Chem. Soc. **110**, 3959 (1988).

34. MURAKAMI, M., H. NAKANO, K. YAMAGUCHI, S. KONOSU, O. NAKAYAMA, Y. MATSUMOTO, and H. IWAMOTO: Meijicoccene, a new cyclic hydrocarbon from *Botryococcus braunii*. Phytochemistry **27**, 455 (1988).

35. DOUGLAS, A.G., K. DOURAGHI-ZADEH, and G. EGLINTON: The fatty acids of the alga *Botryococcus braunii*. Phytochemistry **8**, 285 (1969).

36. DUBREUIL, C., S. DERENNE, C. LARGEAU, C. BERKALOFF and B. ROUSSEAU: Mechanism of formation and chemical structure of Coorongite. Role of the resistant biopolymer and of the hydrocarbons of *Botryococcus braunii*. Ultrastructure of Coorongite and its relationship with Torbanite. Org. Geochem. **14**, 543 (1989).

37. HUANG, Z., and C.D. POULTER: Isoshowacene, a C_{31} hydrocarbon from *Botryococcus braunii* var. Showa. Phytochemistry **28**, 3043 (1989).

38. WHITE, J.D., T.C. SOMERS, and G.N. REDDY: Absolute configuration of ($-$) Botryococcene. J. Am. Chem. Soc. **108**, 5352 (1986).

39. GALBRAITH, M.N., L.W. HILLEN, and L.V. WAKE: Darwinene: a branched hydrocarbon from a green form of *Botryococcus braunii*. Phytochemistry **22**, 1441 (1983).

40. STOILOV, I.L., J.E. THOMPSON, J-H. CHO, and C. DJERASSI: Biosynthetic studies of marine lipids. Stereochemical aspects and hydrogen migrations in the biosynthesis of the triply alkylated side chain of the sponge sterol strongylosterol. J. Am. Chem. Soc. **108**, 8235 (1986).

41. DAVID, M., P. METZGER, and E. CASADEVALL: Two cyclobotryococcenes from the B race of the green alga *Botryococcus braunii*. Phytochemistry **27**, 2863 (1988).

42. JAENICKE, L., and F.-J. MARNER: The irones and their precursors. In: Progress in the Chemistry of Organic Natural Products, vol. 50, ed. by W. HERZ, H. GRISEBACH, G.W. KIRBY, and C. TAMM. Wien, New York Springer. 1986.

43. HUANG, Z., and C.D. POULTER: Braunicene. Absolute stereochemistry of the cyclo-hexane ring. J. Org. Chem. **53**, 4089 (1988).

44. ——: Isobraunicene, Wolficene, and Isowolficene. New cyclic 1'-3 fused isoprenoids from *Botryococcus braunii*. J. Org. Chem. **53**, 5390 (1988).

45. POULTER, C.D.: Biosynthesis of non-head-to-tail terpenes. Formation of 1'-1 and 1'-3 linkages. Acc. Chem. Res. **23**, 70 (1990).

46. HUANG, Z., and C.D. POULTER: Stereochemical studies of botryococcene biosynthesis: analogies between 1'-1 and 1'-3 condensations in the isoprenoid pathway. J. Am. Chem. Soc. **111**, 2713 (1989).

47. CASADEVALL, E., P. METZGER, and M.-P. PUECH: Biosynthesis of triterpenoid hydro-carbons in the alga *Botryococcus braunii*. Tetrahedron Letters **25**, 4123 (1984).

48. WOLF, F.R., E.K. NEMETHY, J.H. BLANDING, and J.A. BASSHAM: Biosynthesis of unusual acyclic isoprenoids in the alga *Botryococcus braunii*. Phytochemistry **24**, 733 (1985).

49. ZUNDEL, M., and M. ROHMER: Procaryotic triterpenoid. 3. The biosynthesis of 2β-methylhopanoids and 3β-methylhopanoids of *Methylobacterium organophilum* and *Acetobacter pasteurianus* ssp *pasteurianus*. Eur. J. Biochem. **150**, 35 (1985).

50. HUANG, Z., and C.D. POULTER: Tetramethylsqualene, a triterpene from *Botryococcus braunii* var. *Showa*. Phytochemistry **28**, 1467 (1989).

51. METZGER, P., and E. CASADEVALL: Structure de trois nouveaux botryococcenes synthétisés par une souche de *Botryococcus braunii* cultivée en laboratoire. Tetrahe-dron Letters **24**, 4013 (1983).

52. WHITE, J.D., G.N. REDDY, and G.O. SPESSARD: Total synthesis of (−)-Botryococcene, J. Am. Chem. Soc. **110**, 1624 (1988).

53. HIRD, N.W., T.V. LEE, A.J. LEIGH, J.R. MAXWELL, and T.M. PEAKMAN: The total synthesis of 10-(R,S)-C$_{30}$ botryococcene and botryococcane and a new synthesis of a general intermediate to the botryococcene family. Tetrahedron Letters **30**, 4867 (1989).

54. GRUNG, M., P. METZGER, and S. LIAAEN-JENSEN: Primary and secondary carotenoids in two races of the green alga *Botryococcus braunii*. Biochem. Syst. Ecol. **17**, 263 (1989).

55. GRUNG, M.: Unpublished results.

56. METZGER, P., and E. CASADEVALL: Aldehydes, very long chain alkenylphenols, epoxides and other lipids from an alkadiene-producing strain of *Botryococcus braunii*. Phytochemistry **28**, 2097 (1989).

57. BRENNAN, J.: *Mycobacterium* and other actinomycetes. In: Microbial lipids, vol. 1, ed. by C. RATLEDGE and S.G. WILKINSON. London: Academic Press. 1988.

58. COJOCARU, M., S. DROBY, E. GLOTTER, A. GOLDMAN, H.E. GOTTLIEB, B. JACOBY, and D. PRUSKY: 5-(12-Heptadecenyl)-resorcinol, the major component of the antifungal activity in the peel of mango fruit. Phytochemistry **25**, 1093 (1986).

59. METZGER, P., and E. CASADEVALL: Botryococcoid ethers. A novel type of ether lipids isolated from the green alga *Botryococcus braunii*. Phytochemistry, in press.

60. LEE, M.S., G.-W. QIN, K. NAKANISHI, and M.G. ZAGORSKI: Biosynthetic studies of brevetoxins, potent neurotoxins produced by the dinoflagellate *Gymnodinium breve*. J. Am. Chem. Soc. **111**, 6234 (1989).

61. METZGER, P., and E. CASADEVALL: In preparation.

62. METZGER, P., E. VILLARREAL-ROSALES, and E. CASADEVALL: Methyl-branched fatty aldehydes and fatty acids in *Botryococcus braunii*. Phytochemistry **30**, 185 (1991).

63. HAN, J., H.W.-S. CHAN, and M. CALVIN: Biosynthesis of alkanes in *Nostoc muscorum*. J. Am. Chem. Soc. **91**, 5156 (1969).

64. RAINWATER, D.L., and P.E. KOLATTUKUDY: Fatty acid biosynthesis in *Mycobacterium tuberculosis* var. *bovis Bacillus Calmette-Guérin*. J. Biol. Chem. **260**, 616 (1985).
65. METZGER, P., R. BISCHOFF, and E. CASADEVALL: In preparation.
66. BERKALOFF, C., E. CASADEVALL, C. LARGEAU, P. METZGER, S. PERACCA, and J. VIRLET: The resistant polymer of the walls of the hydrocarbon-rich alga *Botryococcus*. Phytochemistry **22**, 389 (1983).
67. LARGEAU, C., S. DERENNE, E. CASADEVALL, A. KADOURI, and N. SELLIER: Pyrolysis of immature Torbanite and of the resistant biopolymer (PRB A) isolated from extant alga *Botryococcus braunii*. Mechanism of formation and structure of Torbanite. Org. Geochem. **10**, 1023 (1986).
68. LARGEAU, C., S. DERENNE, E. CASADEVALL, A. KADOURI, and P. METZGER: Formation of *Botryococcus*-derived kerogens. Comparative study of immature Torbanites and of the extant alga *Botryococcus braunii*. Org. Geochem. **6**, 327 (1984).
69. LAUREILLARD, J., C. LARGEAU, and E. CASADEVALL: Oleic acid in the biosynthesis of the resistant biopolymers of *Botryococcus braunii*. Phytochemistry **27**, 2095 (1988).
70. DERENNE, S.: In preparation.
71. KADOURI, A., S. DERENNE, C. LARGEAU, E. CASADEVALL, and C. BERKALOFF: Resistant biopolymer in the outer walls of *Botryococcus braunii*, B race. Phytochemistry **27**, 551 (1988).
72. LAUREILLARD, J., C. LARGEAU, F. WAEGHEMAEKER, and E. CASADEVALL: Biosynthesis of the resistant polymer in the alga *Botryococcus braunii*. Studies on the possible direct precursors. Journal of Natural Products **49**, 794 (1986).
73. DERENNE, S., C. LARGEAU, E. CASADEVALL, and C. BERKALOFF: Occurrence of a resistant biopolymer in the L race of *Botryococcus braunii*. Phytochemistry **28**, 1137 (1989).
74. DERENNE, S., C. LARGEAU, E. CASADEVALL, and N. SELLIER: Direct relationship between the resistant biopolymer and the tetraterpenic hydrocarbon in the lycopadiene race of *Botryococcus braunii*. Phytochemistry **29**, 2187 (1990).
75. DERENNE, S., C. LARGEAU, and E. CASADEVALL: Occurrence of tightly bound isoprenoid acids in an algal, resistant biomacromolecule: possible geochemical implications. Org. Geochem. in press.
76. DURAND, B.: Sedimentary organic matter and kerogen. Definition and quantitative importance of kerogen. In: Kerogen, ed. by B. DURAND. Paris: Editions Technip. 1980.
77. TISSOT, B.P., and D.H. WELTE: Kerogen, composition and classification. In: Petroleum formation and occurrence, ed. by B.P. TISSOT and D.H. WELTE. Berlin Heidelberg New York: Springer. 1978.
78. TEMPERLEY, B.N.: The Boghead controversy and the morphology of the Boghead algae. Trans. R. Soc. Edinburgh **43**, 855 (1936).
79. CORREIA, M., and J. CONNAN: Diagenèse naturelle et diagenèse artificielle de la matière organique à éléments végétaux dominants. In: Advances in Organic Geochemistry 1973, ed. by B. TISSOT and F. BIENNER. Paris: Editions Technip. 73 (1974).
80. DERENNE, S., C. LARGEAU, E. CASADEVALL, and F. LAUPRETRE: Structural analysis of two Torbanites at different evolutionary stages. Investigation of the quantitative reliability of fa determination by 13C CP/MAS n.m.r. Fuel **66**, 1084 (1987).
81. DERENNE, S., C. LARGEAU, E. CASADEVALL, and J. CONNAN: Comparison of Torbanites of various origins and evolutionary stages. Bacterial contribution to their formation. Cause of the lack of botryococcane in bitumens. Org. Geochem. **12**, 43 (1988).
82. ————: Mechanism of formation and chemical structure of Coorongite. II. Structure and origin of the labile fraction. Fate of botryococcenes during early diagenesis. Org. Geochem. **13**, 965 (1988).

83. DERENNE, S., C. LARGEAU, E. CASADEVALL, E. TEGELAAR, and J.W. DE LEEUW: Relationship between algal coals and resistant cell wall biopolymers of extant algae as revealed by Py-GC-MS. Fuel Process. Technol. **20**, 93 (1988).

84. LARGEAU, C., P. BERTRAND, P. FOURMONT, S. DERENNE, and E. CASADEVALL: Etude de trois Torbanites par microspectrofluorimétrie: contribution des différentes fractions constitutives dans la fluorescence totale; corrélations avec la structure chimique; relations avec le degré de maturation. Bull. Soc. Géol. France **8**, 993 (1989).

85. KISTER, J., M. GUILIANO, C. LARGEAU, S. DERENNE, and E. CASADEVALL: Characterization of chemical structure, degree of maturation and oil potential of Torbanites (type 1 kerogens) by quantitative FT-i.r. spectroscopy. Fuel. **69**, 1356 (1990).

86. NIP, M., E.W. TEGELAAR, H. BRINKHUIS, J.W. DE LEEUW, P.A. SCHENCK, and P.J. HOLLOWAY: Analysis of modern and fossil plant cuticles by Curie point Py-GC and Curie point Py-GC-MS: recognition of a new, highly aliphatic and resistant biopolymer. Org. Geochem. **10**, 769 (1986).

87. CHALANSONNET, S., C. LARGEAU, E. CASADEVALL, C. BERKALOFF, G. PENIGUEL, and R. COUDERC: Cyanobacterial resistant biopolymers. Geochemical implications of the properties of *Schizothrix sp.* resistant material. Org. Geochem. **13**, 1003 (1988).

88. ZELIBOR, J.L., L. ROMANKIW, P.G. HATCHER, and R.R. COLWELL: Comparative analysis of the chemical composition of mixed and pure cultures of green algae and their decomposed residues by ^{13}C-NMR spectroscopy. Appl. Environ. Microbiol. **54**, 1051 (1988).

89. GOTH, K., J.W. DE LEEUW, W. PÜTTMANN, and E.W. TEGELAAR: Origin of Messel Oil Shale Kerogen. Nature (London) **336**, 759 (1988).

90. TEGELAAR, E.W., J.W. DE LEEUW, C. LARGEAU, S. DERENNE, H.R. SCHULTEN, R. MÜLLER, J.J. BON, M. NIP, and J.C.M. SPRENKELS: Scope and limitation of several pyrolysis methods in the structural elucidation of a macromolecular plant constituent in the leaf cuticle of *Agave Americana L.* J. Anal. Appl. Pyrolysis **15**, 29 (1989).

91. LARGEAU, C., S. DERENNE, E. CASADEVALL, C. BERKALOFF, M. COROLLEUR, B. LUGARDON, J.F. RAYNAUD, and J. CONNAN: Occurrence and origin of "ultralaminar" structures in "amorphous" kerogens of various source rocks and oil shales. Org. Geochem. **16**, 889 (1990).

92. LARGEAU, C., S. DERENNE, C. CLAIRAY, E. CASADEVALL, J.F. RAYNAUD, B. LUGARDIN, C. BERKALOFF, M. COROLLEUR, and B. ROUSSEAU: Characterization of various kerogens by Scanning Electron Microscopy (SEM) and Transmission Electron Microscopy (TEM). Morphological relationships with resistant outer walls in extant microorganisms. In: Proceedings of the International Symposium on Organic Petrology, Zeist 1990, ed. by W.J.J. FERMONT et J.W. WEEGINK, Geological Survey of the Netherlands, special issue, in press.

93. DERENNE, S., C. LARGEAU, E. CASADEVALL, C. BERKALOFF, and B. ROUSSEAU: Chemical evidence of kerogen formation in source rocks and oil shales *via* selective preservation of thin resistant outer walls of microalgae. Origin of ultralaminae. Geochim. Cosmochim. Acta. In press.

94. TEGELAAR, E.W., J.W. DE LEEUW, S. DERENNE, and C. LARGEAU: A reappraisal of kerogen formation. Geochim. Cosmochim. Acta **53**, 3103 (1989).

95. DE LEEUW, J.W., and C. LARGEAU: A review of macromolecular organic compounds that comprise living organisms and their role in kerogen, coal and petroleum formation. In Organic Geochemistry, ed. by M.H. ENGEL and S.A. MACKO. New York: Plenum Publishing Corp. In press.

96. SEIFFERT, W.K., and J.M. MOLDOWAN: Paleoreconstruction by biological markers. Geochim. Cosmochim. Acta **45**, 783 (1981).

97. TULLOCH, A.P.: Chemistry of waxes of higher plants. In: Chemistry and biochemistry of natural waxes, ed. by P.E. KOLATTUKUDY. Amsterdam: Elsevier. 1976.

98. MOLDOWAN, J.M., and W.K. SEIFERT: First discovery of botryococcane in petroleum. J.C.S. Chem. Commun. **19**, 912 (1980).

99. MCKIRDY, D.M., R.E. COX, J.K. VOLKMAN, and V.J. HOWELL: Botryococcane in a new class of Australian non-marine crude oils. Nature (London) **320**, 57 (1986).

100. BRASSELL, S.C., G. EGLINTON, and F. JIA MO: Biological markers compounds as indicators of the depositional history of the Maoming Oil Shale. Org. Geochem. **10**, 927 (1986).

101. CHU, S.P.: The influence of the mineral composition of the medium on the growth of planktonic algae. I. Methods and culture media. J. Ecol. **30**, 284 (1942).

102. BELCHER, J.H.: Notes on the physiology of *Botryococcus braunii* Kützing. Arch. Mikrobiol. **61**, 335 (1968).

103. LARGEAU, C., E. CASADEVALL, and D. DIF: Renewable hydrocarbon production from the alga *Botryococcus braunii*. In: Energy from biomass. 1st E.C. Conference, ed. by W. PALZ, P. CHARTIER, and D.O. HALL. London: Applied Science Publishers. 1981.

104. CASADEVALL, E., D. DIF, C. LARGEAU, C. GUDIN, D. CHAUMONT, and O. DESANTI: Studies on batch and continuous cultures of *Botryococcus braunii*: hydrocarbon production in relation to physiological state, cell ultrastructure and phosphate nutrition. Biotechnol. Bioengin. **27**, 286 (1985).

105. METZGER, P., E. CASADEVALL, A. COUTE, and Y. POUET: Screening of wild strains of the hydrocarbon-rich alga *Botryococcus braunii*. Productivity and hydrocarbon nature. In: Energy from biomass. 3rd E.C. Conference, ed. by W. PALZ, J. COOMBS, and D.O. HALL. London: Elsevier. 1985.

106. WEETALL, H.: Studies on the nutritional requirements of the oil-producing alga *Botryococcus braunii*. Appl. Biochem. Biotechnol. **11**, 377 (1985).

107. TENAUD, M., M. OHMORI, and S. MIYACHI: Inorganic carbon and acetate assimilation in *Botryococcus braunii* (Chlorophyceae). J. Phycol. **25**, 662 (1989).

108. IWAMOTO, H., and A. SUZUKI: Fat synthesis in unicellular algae. Part II: Chemical composition of nitrogen deficient *Chlorella* cells. Bull. Agr. Chem. Soc. **19**, 247 (1955).

109. AARONSON, S., T. BERNER, and Z. DUBINSKY: Microalgae as a source of chemicals and natural products. In: Algae biomass, production and use, ed. by G. SHELEF and C.J. SOEDER. Amsterdam: Elsevier. 1980.

110. SHIFRIN, N.S., and S.W. CHISHOLM: Phytoplankton lipids: Environmental influence on production and possible commercial applications. In: Algae biomass, production and use, ed. by G. SHELEF and C.J. SOEDER. Amsterdam: Elsevier. 1980.

111. BRENCKMANN, F., C. LARGEAU, E. CASADEVALL, and C. BERKALOFF: Influence de la nutrition azotée sur la croissance et la production d'hydrocarbures de l'algue uni-cellulaire *Botryococcus braunii*. In: Energy from biomass, 3rd E.C. Conference, ed. by W. PALZ, J. COOMBS, and D.O. HALL. London: Elsevier. 1985.

112. DUBINSKY, Z., T. BERNER, and S. AARONSON: Potential of large-scale algal cultures for biomass and lipid production in arid lands. Biotechnol. Bioengin. Symp. N°8, 51 (1978).

113. BRENCKMANN, F., C. LARGEAU, E. CASADEVALL, B. CORRE, and C. BERKALOFF: Influence of light intensity on hydrocarbon and total biomass production of *Botryococcus braunii*. Relationships with photosynthetic characteristics. In: Energy from biomass. 3rd E.C. Conference, ed. by W. PALZ, J. COOMBS, and D.O. HALL. London: Elsevier. 1985.

114. CHIRAC, C., E. CASADEVALL, C. LARGEAU, and P. METZGER: Influence de la souche et

des bactéries associées sur la productivité en hydrocarbures de l'algue *Botryococcus braunii*. C.R. Acad. Sci. Paris **295 III**, 671 (1982).

115. CHIRAC, C., E. CASADEVALL, and C. LARGEAU: Croissance et production d'hydrocarbures de l'algue *Botryococcus braunii* en cultures associées. C.R. Acad. Sci. Paris **297 III**, 187 (1983).

116. CHIRAC, C., E. CASADEVALL, C. LARGEAU, and P. METZGER: Bacterial influence upon growth and hydrocarbon production of the green alga *Botryococcus braunii*. J. Phycol. **21**, 380 (1985).

117. BAILLIEZ, C., C. LARGEAU, E. CASADEVALL, and C. BERKALOFF: Effets de l'immobilisation en gel d'alginate sur l'algue *Botryococcus braunii*. C.R. Acad. Sci. Paris **296 III**, 199 (1983).

118. BAILLIEZ, C., C. LARGEAU, and E. CASADEVALL: Effect of immobilization on the hydrocarbon-rich alga *Botryococcus braunii*. In: Energy from biomass. 2nd E.C. Conference, ed. by A. STRUB, P. CHARTIER, and G. SCHLESER. London: Applied Sciences Publishers. 1983.

119. BAILLIEZ, C., C. LARGEAU, and E. CASADEVALL: Growth and hydrocarbon production of *Botryococcus braunii* immobilized in calcium alginate gel. Appl. Microbiol. Biotechnol. **23**, 99 (1985).

120. BAILLIEZ, C., C. LARGEAU, C. BERKALOFF, and E. CASADEVALL: Immobilization of *Botryococcus braunii* in alginate: Influence on chlorophyll content, photosynthetic activity and degeneration during batch cultures. Appl. Microbiol. Biotechnol. **23**, 361 (1986).

121. BAILLIEZ, C., C. LARGEAU, E. CASADEVALL, L.W. YANG, and C. BERKALOFF: Photosynthesis, growth and hydrocarbon production of *Botryococcus braunii* immobilized by entrapment and adsorption in polyurethane foams. Appl. Microbiol. Biotechnol. **29**, 141 (1988).

122. FRENZ, J., C. LARGEAU, E. CASADEVALL, F. KOLLERUP, and A.J. DAUGULIS: Hydrocarbon recovery and biocompatibility of solvents for extractions from cultures of *Botryococcus braunii*. Biotechnol. Bioengin. **34**, 755 (1989).

123. FRENZ, J., C. LARGEAU, and E. CASADEVALL: Hydrocarbon recovery by extraction with a biocompatible solvent from free and immobilized cultures of *Botryococcus braunii*. Enzyme Microb. Technol. **11**, 717 (1989).

(*Received December 12, 1990*)

Carbazole Alkaloids III*

D. P. CHAKRABORTY** and SHYAMALI ROY***

Comparative Phytochemical Laboratory, Bose Institute, Calcutta, India

Contents

* Part I—Ref. (18); Part II—Ref. (11).
** Present address: 11/1/5, Satchasipara Lane, Calcutta 700036, India.
*** Present address: 18, J. N. Roy Lane, Calcutta 700006, India.

I. Introduction

Major developments in the chemistry of carbazoles until 1960 arose mainly as a result of their use in the dye stuff and polymer industries although carbazole itself (1) was discovered in coal tar in 1872 by GRAEBE and GLAZER (53). Since the report of the first carbazole alkaloid murrayanine (2) from *Murraya koeinigii* Spreng and the antibiotic properties of these alkaloids in 1965 by CHAKRABORTY *et al.* (22, 31), there has been wider interest on the structure, synthesis and biochemical properties of these compounds. At present more than one hundred alkaloids are known and various biochemical and medicinal properties of this group of compounds have been investigated. Several reviews relating to carbazole alkaloids have appeared since 1971 (67, 18, 19, 20, 56, 11). The present review relates to work reported after the previous review (11) and earlier results not included therein.

A. Nomenclature

The nomenclature adopted in this review for the tricyclic system is the same as that used by Chemical Abstracts (*94*). The rings of the conventional tricyclic system are denoted by A, B, C. Abbreviations like C-1 and H-1 are used to denote the carbon and the hydrogen at a particular position.

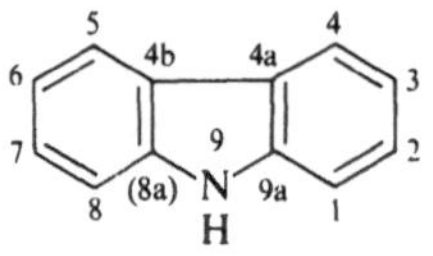

(1) 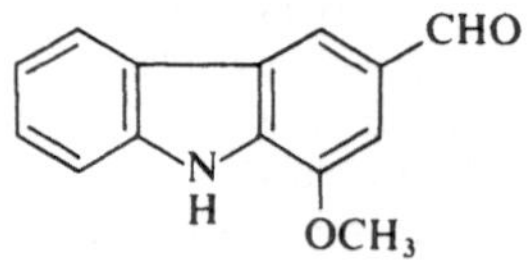   (2)

B. Occurrence

Carbazole alkaloids are abundant in higher plants belonging to the subtribe Clausenae, subfamily Aurantoidae of the family Rutaceae. The genera *Micromelum* of Rutaceae and *Ekebergia* of Meliaceae have been reported to elaborate carbazole alkaloids. Carbazole alkaloids have also been reported from lower plants which include *Streptoverticillium ehiminse*, *Streptomyces murayamaensis* (Actinomycetes group), the blue green alga Hyella as well as from mammalian sources like bovine urine.

The genus Murraya, specially a species native to Taiwan (*M. euchrestifolia*) is the richest source of carbazole alkaloids (see Table 1). Various monomeric alkaloids (C_{13}-, C_{18}-, C_{23}-skeletons or their derivatives) and bis-alkaloids built on these monomeric units have been reported from *M. euchrestifolia* (*81*).

Table 1. *Distribution of Carbazole Alkaloids**

	Name of the Alkaloids and Sources
Plant group/type	Figures in the bracket represent the plant source specified here: 1 *Murraya koenigii* Spreng., 2. *M. paniculata* L. Jack (Syn. *M. exotica* Linn.), 3. *M. euchrestifolia* Hyata., 4. *Glycosmis pentaphylla* (Retz) D.C., 5. *Clausena anisata* (Willd) Olive., 6. *C. heptaphylla* Wt. & Arn., 7. *C. indica* Olive., 8. *C. lansium* (Lour.) Skeels (Syn. *C. wampi* Olive.), 9. *Micromelum zelynicum,* 10. *Ekebergia senegalensis.*

Table 1 (*continued*)

Name of the Alkaloids and Sources

A. *Higher Plants*
(i) C_{13}-
    skeleton

Murrayafoline A (1, 3); 2-hydroxy-3-methylcarbazole (1); 2-methoxy-3-methylcarbazole (1); koeniline (1); mukoline (1); murrayanine (1, 2, 3, 6); mukolidine (1); mukonal (1); mukoic acid (1, 3); mukonine (1); mukonidine (1); 3-methylcarbazole (2, 3, 4, 6); murrayastine (3); murrayaline (3); murrayaquinone A (3); glycozoline (4); glycozolidine (4); glycozolidol (4) glycozolidal (4); 3-Formylcarbazole (3); N-Methoxy-3-formylcarbazole (3); carbazole (4); O-demethylmurrayanine (5).

(ii) C_{18}-
     skeleton

Girinimbine (1, 2, 6); koenine (1); koenimbine (1); koenigine (1); koenigicine, koenidine (1, 9); mukonicine (1); murrayacine (1, 6); dihydroxygirinimbine (3); murrayafoline B (3); pyrayafoline (3); murrayaquinone B (3); mupamine (4, 5); clausanitin (5); clausenapin (6); heptaphylline (6); heptazoline (6); heptazolidine (6); heptazolicine (6); 6-methoxyheptaphylline (7); indizoline (7); lansine (8); ekebergenine (10); isomurrayafoline B (3); atanisatin (5); 3-methyl6:5-(2'2'-dimethyl-$\Delta^{3'}$)-pyranocarbazole (3); 5:6 Furano-3-methylcarbazole (3)

(iii) C_{23}-
      skeleton

Mahanimbiol (1); mahanimbine (1); dl-mahanimbine (1, 3); isomahanimbine/mahanimbicine (1); mahanine (1); mahanimbine (1); mahanimboline (1); ($-$) murrayazoline (1); ($\pm$) murrayazoline/mahanimbidine/currayagine (1); ($+$) murrayazoline (3); isomurrayazoline (1); murrayazolidine (1); cyclomahanimbine/-dl-murrayazolidine (1,3); currayanine (1); bicyclomahanimbine (1); bicyclomahanimbicine (1); exozoline (2); murrayaquinone C (3); murrayaquinone D (3); murrayazolinol (1).

(iv) *Dimeric*
     carbazoles

Bismurrayazoline A (3); bismurrayafoline B (3); murrafoline (3); murrafoline B, C, D, E, F (3); bismurrayafolinol (3); oxydimurrayafoline (3); N-dihydropyranocarbazole-murrayafolinol (3) (60)

B. *Lower Plants*
 (i) *Marine Source*
     (Cyanophyta)

Hyellazole, 6-chlorohyellazole (*Hyella caespitosa* Bron et Flah) (Cyanophyta)

 (ii) *Microbial*
      Source
      (Schizomicophyta)

Carbazomycin A, B, C, D, G, H (*Streptoverticillum ehimense* H 1051-NY10); carbazomycinal (carbazomycine E); 6-methoxy-carbazomycinal (carbazomycine F) (*Streptomyces* Sp.); kinamycin A, B, C, D, E, F, prekinamycin, ketoanhydrokinamycin, (*S. murayamaensis*); tubingensin A, Tubingensin B (*Aspergillus tubingensis*).

Carbazoles from
other sources

3-Chlorocarbazole (bovine urine); carbazole (coal tar); methylcarbazoles (Kuwait petroleum) (20)

* For structures consult references (*11, 18, 19*) besides the present review.

C. Detection of Carbazoles

Newer chromatographic methods have been used for the detection and separation of carbazoles (*4, 34, 57, 96*) from various sources.

Pyranocarbazoles like girinimbine (**3**) can be detected readily by the formation of an immediate blue colour with BF$_3$-etherate (*16*). Pyrano-coumarins like seselin (**4**) (*32*) do not respond to this test.

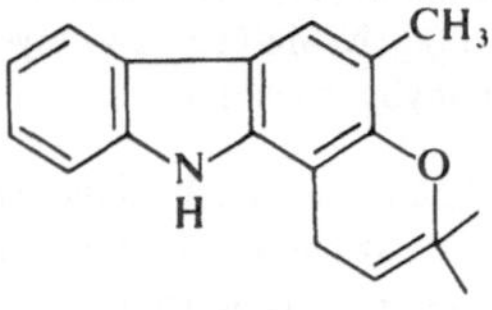

(3) 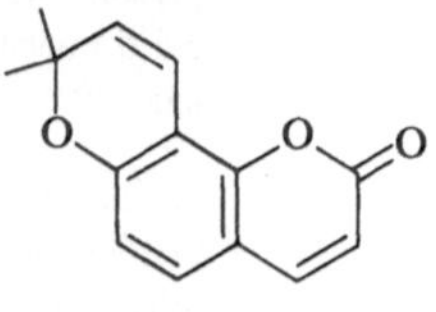   (4)

II. Methods of Structure Elucidation

A. Physical Methods

Significant developments relating to the application of the physical methods for structure determination of carbazole alkaloids after 1977 (*18*) are briefly summarised here.

1. Ultraviolet Absorption Spectra

The characteristic ultraviolet absorption spectra of carbazole, 3-methyl-, formyl- and methoxycarbazoles (*18*) continue to be useful in structure elucidation studies. Uv absorption data of the alkaloids reported after 1977 are detailed in Table 2.

2. IR Spectra

A strong IR band near 750 cm^{-1} is characteristic of the four adjacent unsubstituted positions in ring A of carbazomycin-B (**5**) and -A (**6**) (*107*), while a carbonyl bond at the unusually low frequency of 1630–1645 cm^{-1} is diagnostic of a formyl group at C-3 chelated to the hydroxyl at C-2 in various alkaloids.

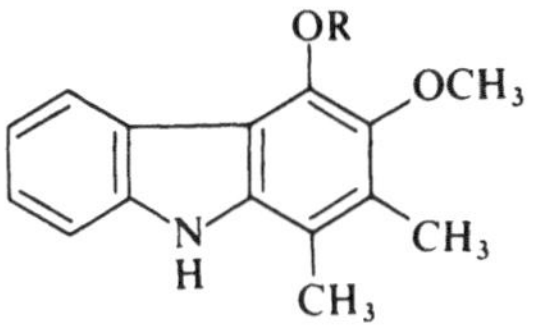

(5) R = H
(6) R = CH₃

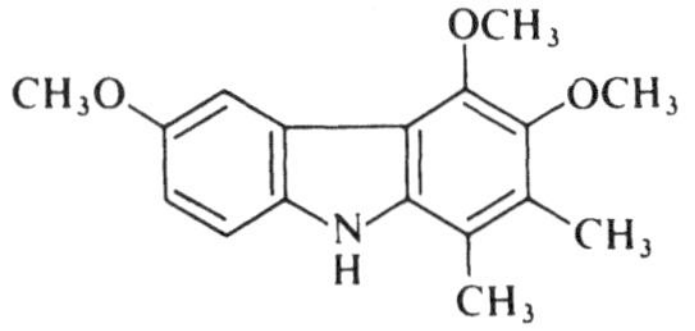

(7)

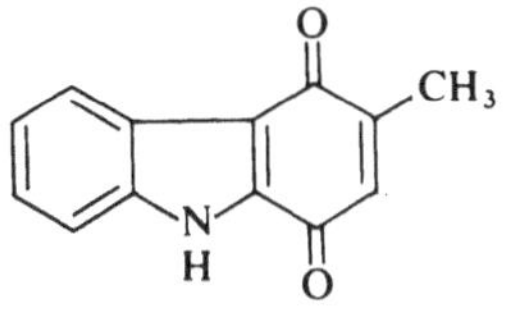

(8)

3. NMR Spectra

(a) *¹H-NMR Spectra*: The low field H-5 and H-4 signals (δ 7.17–δ 7.50), coupling with *ortho-* or *meta* protons and shifts due to vicinal functional groups are useful for the detection of substituents in rings A and C. The lack of substitution in ring A has been inferred from the presence of an isolated four spin system (*43*) in the region δ 7.06 to δ 7.90. The anisotropic deshielding effect of ring A on the methoxy group at C-4 has been recorded (δ 4.06) in carbazomycin D (7). The quinonoid keto group at C-4 in carbazoloquinones like murrayaquinone A (8) has been found to deshield the H-5 signal (*11*).

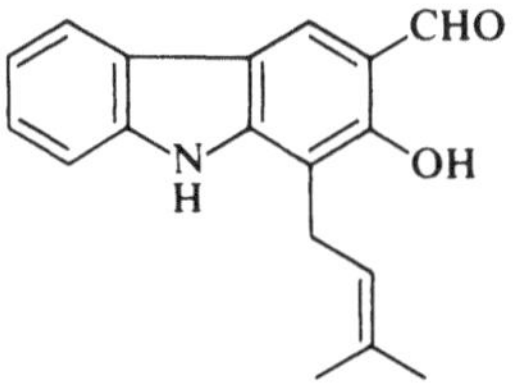

(9) 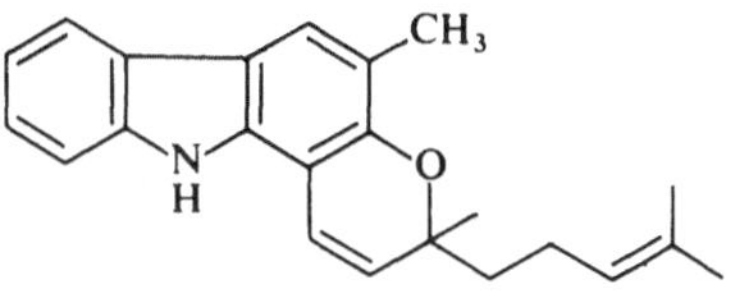   (10)

The signals of the benzylic protons (δ 3.6), the gem-dimethyl group (1.66, 1.83) and the vinylic proton (δ 5.35) as, for example, in heptaphylline (9) are diagnostic for the 3:3 dimethyl allyl side chain (abbreviated as DMA). The doublets for each of the olefinic protons (*J* = 10 Hz) at δ 6.25 and 5.4 respectively together with the signal for the *gem*-dimethyl group on the carbon linked to the oxygen are indicative for the

Table 2. *Ultraviolet Absorption Spectra of Some Carbazole Alkaloids*
(For structures not numbered see Refs *11, 19*)

Compound	λ_{max} in nm	Log ε/ε
1. Murrayafoline-A (12)	225, 243, 254(sh), 283(sh), 292, 330, 344	4.47, 4.58, 4.44, 3.83, 4.01, 3.53, 3.49
2. Murrayastine (13)	224, 247, 255, 298, 322, 336	— — —
3. 1-Hydroxy-3-methyl-carbazole	224, 242, 292, 330	4.59, 4.69, 4.09, 3.58
4. 2-Hydroxy-3-methyl-carbazole	235, 254, 258, 304, 332	4.65, 4.25, 4.26, 4.19, 3.66
5. 2-Methoxy-3-methyl carbazole	235, 255, 300, 328	4.35, 3.8, 3.9, 3.30
6. 3-Formylcarbazole (218)	212, 233, 244(sh), 273 288, 325	— — —
7. N-Methoxy-3-formyl-carbazole (219)	236, 272, 288, 320	— — — — —
8. O-Demethyl-murrayanine (217)	226, 244, 255, 278, 291, 336, 346	4.40, 4.51, 4.39, 4.59, 4.45, 4.22, 4.22
9. Glycozolinine (Glycozolinol)	225, 255, 270, 302	4.30, 4.02, 3.9, 4.2
10. Glycozolidol	232, 260, 305	4.50, 4.10, 4.29
11. Murrayaline (14)	226, 260(sh), 304, 383	— — —
12. Koeniline	225, 241, 251, 258, 279, 289, 323, 335	4.56, 4.71, 4.22, 4.38, 3.86, 4.04, 3.61, 3.58
13. Mukoline	221, 242, 252, 258, 280, 290, 320	4.60, 4.85, 4.65, 4.0, 3.4, 3.6, 3.0
14. Mukolidine	238, 248, 275, 290, 330	4.74, 4.46, 4.82, 4.48, 4.34
15. Mukonal	234, 247, 278, 297, 342	4.42, 4.21, 4.54, 4.58, 4.06
16. Lansine	233, 248, 309, 340, 368	— — —
17. Glycozolidal	235, 248, 303, 340	4.4, 4.2, 4.2, 4.06
18. Murrayaquinone A	225, 258, 293(sh), 398	4.63, 4.51, 3.85, 3.93
19. Clausenapin	224, 240, 250, 290	4.55, 4.59, 4.50, 4.29
20. Murrayafoline B	230(sh), 247, 255(sh), 302, 324(sh), 337(sh)	— — —
21. Isomurrayafoline B (223)	213, 237, 264, 310, 330(sh)	— — —
22. Clausanitin	239, 249, 278, 288, 297, 340	4.40, 4.29, 4.39, 4.46, 4.53, 3.90
23. Mupamine	238, 272, 282, 320	4.89, 4.09, 4.45, 3.43
24. Mukonicine	226, 240, 300, 342	4.70, 4.67, 4.59, 4.26
25. Pyrayafoline (121)	222, 239, 286(sh), 295, 334	— — —
26. Heptazolicine	242, 275, 300	4.65, 4.62, 4.40
27. Dihydroxygirinimbine (220)	215, 238, 254, 259, 303, 332	— — —
28. Murrayaquinone B	210(sh), 231, 264, 310(sh) 404	4.28, 4.58, 4.44, 3.21, 3.66
29. Pyrayaquinone A (25)	220(sh), 252, 308(sh), 460	— — —
30. Pyrayaquinone B (26)	229(sh), 248, 295(sh), 320, 410	— — —
31. Mahanimbinol	244, 260, 296, 325, 339	4.26, 4.14, 3.98, 3.40, 3.38

Table 2 (*continued*)
(For structures not numbered see Refs *11*, *19*)

Compound	λ_{max} in nm	Log ε/ε
32. Mahanimboline	237, 280, 288, 330	4.42, 3.69, 4.25, 2.75
33. (+) Murrayazoline	223(sh), 246, 261(sh), 309, 341	4.39, 4.66, 4.29, 4.14, 3.66
34. Exozoline	242, 256, 306	4.47, 4.18, 3.97
35. Murrayazolinol (**227**)	245, 265, 305	4.85, 4.32, 4.40
36. Isomurrayazoline	240, 253, 258, 303, 330	4.65, 4.15, 4.08, 3.85, 3.47
37. Pyrayaquinone C (**27**)	248, 275(sh), 294, 3.92	— — —
38. Murrayaquinone C (**17**)	233, 267, 410	— — —
39. Murrayaquinone D (**16**)	215(sh), 234, 266, 415	— — —
40. Bismurrayafoline A (**35**)	228, 244, 253(sh), 284(sh), 283, 340	4.81, 4.95, 4.84, 4.15, 4.30, 3.92
41. Bismurrayafoline B (**36**)	225(Sh), 240, 265(Sh), 285(Sh), 312, 333(Sh)	— — —
42. Bismurrayafolinol (**37**)	225, 244, 252, 281, 292, 329, 340	— — —
43. Oxydimurrayafoline (**38**)	226, 242, 253, 260, 280, 291, 324, 337	— — —
44. Murrafoline B (**19**)	208, 226, 240, 292, 304, 330	— — —
45. Murrafoline C (**34**)	224, 242, 251, 291, 328, 342	— — —
46. Murrafoline D (**157**)	228, 242, 293, 302, 332, 346	— — —
47. Murrafoline E (**229**)	228(sh), 238, 255(sh), 263(sh), 287, 328, 340, 352	— — —
48. Murrafoline F (**228**)	218(sh), 239, 252(sh), 261(sh), 285(sh), 296, 326, 340	— — —
49. Carbazomycin A (**6**)	223, 242, 293, 324, 340	36000, 51000, 20700, 4700, 4600
50. Carbazomycin B (**5**)	224, 245, 289, 330, 340	38000, 46800, 16100, 5300, 5800
51. Carbazomycin C (**39**)	227, 245, 260(sh), 287(sh), 295, 341, 354	24900, 24000, 12500, 7600, 12200, 3600, 4200
52. Carbazomycin D (**7**)	229, 247, 255(sh), 291(sh), 300, 340, 356	30500, 27800, 19300, 10300, 17000, 4300, 4500
53. Carbazomycinal (**11**) (Carbazomycin E)	214, 227, 263, 295, 320, 372	24400, 23200, 12200, 16300, 5100, 8400
54. 6-Methoxycarbazomycinal (**24**)	215, 227, 245, 268, 310, 382	27000, 24500, 13000, 13000, 17000, 8500
55. Carbazomycin G (**39**)	214, 253, 272(sh), 278, 340	33200, 19800, 8600, 7400, 5600
56. Carbazomycin H (**230**)	214, 253, 261(sh), 292, 310(sh) 340	35700, 23800, 20300, 10100, 5800, 3500
57. Kinamycin A (**193**)	248, 279	— — —
58. Kinamycin B (**195**)	248, 277(sh), 300(sh)	— — —
59. Kinamycin C (**42**)	246, 275, 370	— — —
60. Kinamycin D (**194**)	257, 278(sh)	— — —

Table 2 (*continued*)

Compound	λ_{max} in nm	Log ε/ε
61. Prekinamycin (**190**)	254, 288.4, 342(sh)	5500, 21700, 5770
62. Keto-anhydro kina- mycin (**186**)	235.8, 255, 305.6	43300, 2800, 18600
63. Kinamycin E (**237**)	255.6, 277.0, 295.0	21700, 14100, 9220
64. Tubingensin A (**20**)	218, 239, 262, 302, 326, 340	14900, 18200, 6930, 6780, 480, 480
65. Hyellazole	226(sh), 232, 238(sh), 250(sh), 260(sh), 292(sh), 304, 338, 352	34000, 36500, 34500, 20000, 13500, 12500, 18500, 4500, 5000
66. 6-Chlorohyellazole	220, 235, 242, 255(sh), 272(sh), 300(sh), 310, 348, 360	32600, 32600, 32700, 20000, 14600, 13200, 18000, 4400, 4200
67. 3-Chlorocarbazole (**212**)	215, 223, 229, 236, 246, 260, 291(sh), 297, 319(sh), 330, 343	2030, 2010, 2075, 2200, 1230, 1120, 600, 865, 135, 180, 150

$(-)\varepsilon/\log \varepsilon$ not mentioned in original publication.

$2':2'$-dimethyl-$\Delta^{3'}$ pyran system (abbreviated as DMP) fused to a carbazole skeleton as in girinimbine (**3**). In compounds of the mahanimbine
(**10**) type one of the methyls of a DMP system is replaced by a chain of 6-
carbons, producing the expected changes in the spectrum. The mutually
chelated hydroxyl and formyl groups exhibit signals of the aldehydic
proton at δ 9.30 and the hydroxyl beyond δ 10.00. In Table 3, ^{1}H NMR
data of the alkaloids reported after 1977 (*18*) are briefly summarised (the
signals of the NH proton are not included).

   (b) *Application of the Nuclear Overhauser Effect (NOE)*:
Enhancement of proton signals due to irradiation of methyl and methoxy
protons in NOE experiments has been utilized for locating protons on
ring A and C of the alkaloids (**5, 11–19**). This has been discussed in
Section V for the individual alkaloids.

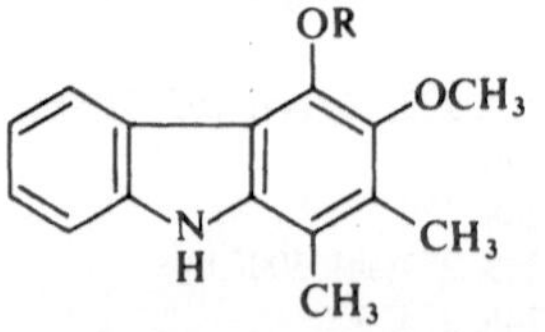

(**5**) R = H

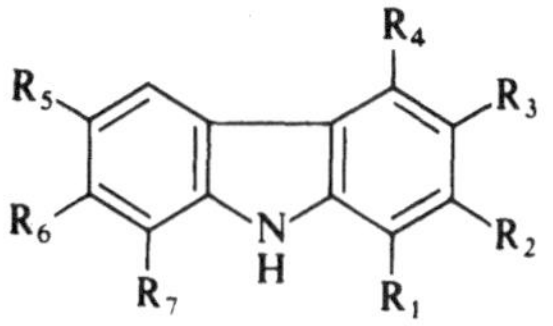

(11) Carbazomycinal R_1 = CHO; R_2 = CH_3
R_3 = OCH_3; R_4 = OH; R_5 = R_6 = R_7 = H
(12) Murrayafoline A R_1 = OCH_3; R_3 = CH_3
R_2 = R_4 = R_5 = R_6 = R_7 = H
(13) Murrayastine R_1 = R_6 = R_7 = OCH_3
R_3 = CH_3; R_2 = R_4 = R_5 = H
(14) Murrayaline R_1 = R_4 = R_5 = H; R_3 = CH_3
R_2 = R_6 = OCH_3; R_7 = CHO
(18) Deoxycarbazomycin R_1 = R_2 = CH_3, R_3 = OCH_3
R_4 = R_5 = R_6 = R_7 = H

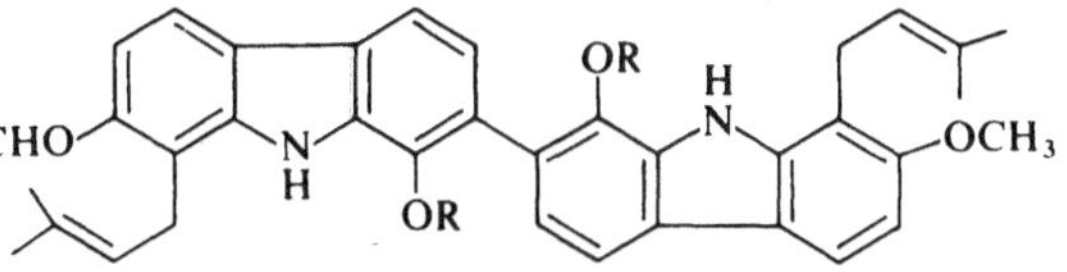

(15) R = $COCH_3$ (Bismurrayafoline B diacetate)

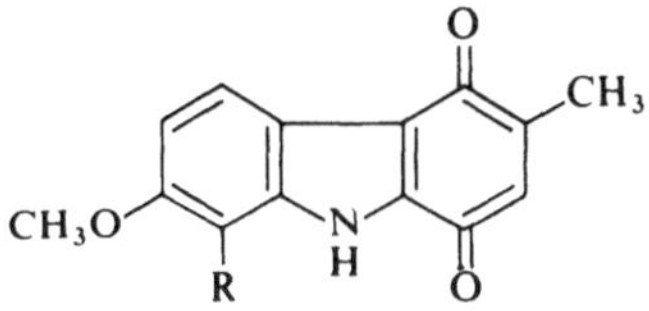

(16) R = geranyl chain
(17) R = DMA

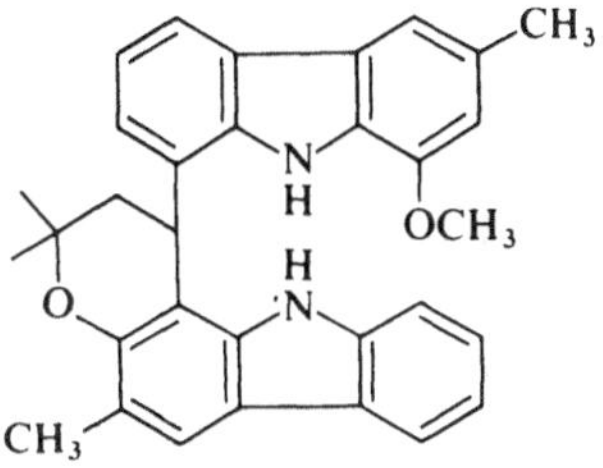

(19) Murrafoline B.

In the case of some complex compounds like tubingensin A and B (20 and 21) the more precise INEPT method has been utilised (*119, 120*).

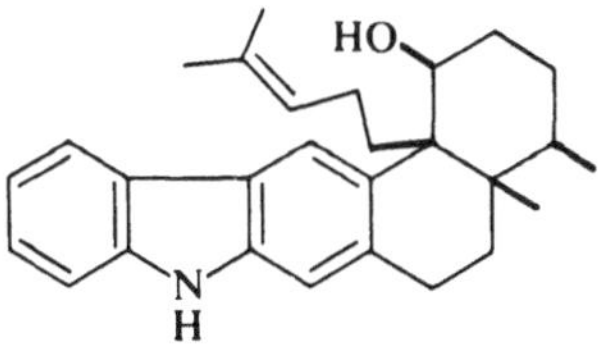

(20)

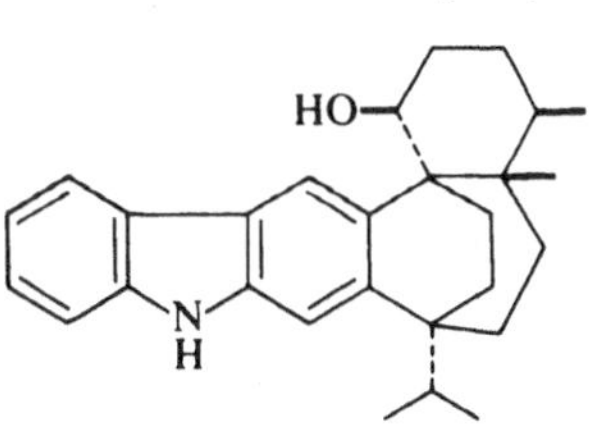

(21)

Table 3. *¹H-NMR Data of Some Carbazoles (δ values)*

Compound	Aromatic protons		Others	Protons of groups on the aromatic ring and other protons
	H-4	H-5		
1. Murrayafoline A (12)	7.33 (s)	7.87 (d, 8 Hz)	6.55 (s), 6.9–7.3 (m)	2.42 (ArM, s); 3.76 (ArOM, s)
2. Murrayastine (13)	7.32 (s)	7.56	6.65 (H-2), 6.80 (H-6)	2.47 (ArM, s); 3.95, 3.96, 4.00 (3 ArOM, s)
3. 2-Hydroxy-3-methylcarbazole	7.68 (s)	—	8.0–7.1 (m); 7.0 (s)	2.33 (ArM, s); 8.1 (ArOH, s)
4. 2-Methoxy-3-methylcarbazole	7.5 (s)	—	7.4–7.2 (m); 6.95 (s)	2.35 (ArM, s); 3.77 (ArOM, s)
5. 3-Formylcarbazole (218)	8.73 (d, 1.7 Hz)	8.29 (d)	7.61 (d), 7.49, 7.3 (t, each 7.7 Hz); 7.98 (dd, 8.4 & 1.7 Hz)	
6. O-Demethylmurrayanine (217)	8.15 (d, 1.4 Hz)	8.08 (ddd, 7.8, 1.2, 0.8 Hz)	7.33 (d, 1.4 Hz, H-2); 7.21, (ddd, 7.8, 7.0, 1.2 Hz, H-6), 7.41 (ddd, 8.2, 7.0, 1.2 Hz, H-7); 7.53 (ddd, 8.2, 1.2, 0.8 Hz, H-8)	9.89 (ArCHO, s)
7. N-Methoxy-3-formyl-carbazole (219)	8.73 (d, 1.4 Hz)	8.31 (d, 7.7 Hz)	8.08 (dd, 8.4, 1.4 Hz); 7.75 (d, 8.4 Hz); 7.61 (d, 7.7 Hz); 7.60 (t, 7.7 Hz); 7.37 (t, 7.7 Hz)	10.12 (ArCHO, s), 4.27 (Ar-OM, s)
8. Glycozolinine/Glycozolinol	7.8 (d)	7.8 (d)	7.20–6.86 (m)	2.5 (ArM, s); 11.1 (ArOH, s)
9. Glycozolidol	7.45 (s)	7.2 (d)	6.92 (s, H-1); 7.0 (d, H-8); 6.75 (dd, H-7)	2.38 (ArM, s); 3.7 (ArOM, s); 10.8 (ArOH, s)
10. Murrayaline (14)	7.95 (s)	8.02 (d, 9 Hz)	6.77 (H-6); 6.91 (H-1)	2.35 (ArM, s); 3.90, 3.98 (2 ArOM, s); 10.58 (ArCHO, s)

11. Koeniline	7.66 (br)	8.04 (d, 8.0 Hz)	6.95 (br, H-2); 7.23 (ddd, 8.0, 7.0, 1.3 Hz, H-6); 7.42 (ddd, 7.0, 1.2 Hz, H-7); 7.46 (d, 8.0 Hz, H-8)	4.01 (ArOM); 4.84 (Ar-CH_2OH)
12. Mukonal	8.4 (s)	—	8.0–7.3 (m); 7.0 (s)	10.16 (ArCHO, s); 11.76 (ArOH, s)
13. Mukoline	—	—	7.0–7.9 (m)	4.75 (ArCH_2OH); 4.5 (ArCH_2OH); 3.9 (ArOM, s)
14. Mukolidine	8.08	8.15	7.3–7.6 (complex m)	10.8 (ArCHO, s); 4.05 (ArOM, s)
15. Lansine	8.36 (s)	7.61 (d, 2.0 Hz)	6.91 (dd, 2.0, 9.0 Hz, H-7; 7.30 (d, 9.0 Hz, H-8); 6.80 (s, H-1)	3.86 (ArOM, s); 9.91 (ArCHO, s)
16. Glycozolidal	8.4 (s)	7.6 (d)	6.85 (dd); 7.3 (d)	3.7, 3.9 (2 ArOM, s); 9.9 (ArCHO, s)
17. Murrayaquinone A	—	8.23 (dt, 1 & 5 Hz)	7.30–7.60 (m, H-6, H-7, H-8)	2.19 (Allyl methyl, d, 1.5 Hz); 6.51 (olefinic, 1.5 Hz, H-2)
18. Clausenapin	7.5 (s)	7.9 (q, 7.0, 2.0 Hz);	7.4–7.1 (m)	2.28 (ArM, s); [1.85 & 1.75 (2 Me, s), 3.62 (d, 8 Hz), 5.3 (t, 8 Hz)]-DMA
19. Murrayafoline B	7.32 (s)	7.72 (d, 9 Hz)	6.56 (s, H-2); 6.82 (d, 9 Hz, H-6)	3.90 (ArOM, s); [1.72 & 1.88 (2 Me, s), 3.60 (d, 8 Hz), 5.30 (vinylic, t, 8 Hz)]-DMA
20. Isomurrayafoline B (223)	7.67 (s)	7.70 (d, 8 Hz)	6.82 (d, 8 Hz)	2.38 (ArM, s); 4.74 (ArOH, br, s); 3.9 (ArOM, s); [1.74 (Me, s), 1.88 (Me, s), 3.60 (d, 7 Hz), 5.30 (t, 7 Hz)]-DMA
21. Methyl ekeberginine	—	8.10 (ddd, 8.1, 1.2, 0.7 Hz)	7.43 (s, H-2); 7.30 (H-6); 7.51 (H-7); 7.41 (H-8)	3.99 (ArOM, s); 4.13 (N-Me, s); 10.37 (ArCHO, s) [1.70, 1.89 (2 Me), 4.19 (bz-CH_2) 5.28]-DMA
22. Clausanitin	8.0 (s)	7.9 (d, 7 Hz)	7.34 (H-1); 7.37 (H-8); 7.31 (dd, 7 & 2 Hz, H-6)	9.81 (ArCHO, s) [3.67 (Bz-CH_2, d, 7 Hz); 1.85 & 1.72 (2 Me, s) & 5.33 (m)]-DMA
23. Mukonicine	7.56 (s)	7.40	6.88	3.9 (2ArOM, s); 2.3 (ArM, s) [1.44 (2 Me, s), 6.55, 5.65 (d, 9 Hz each)]-DMP
24. Pyrayafoline (121)	7.62 (s)	7.63 (d, 8 Hz)	6.67 (d, 8 Hz, H-5); 6.83 (s. H-6)	2.33 (ArM, s); 3.89 (ArOM, s) [1.47 (2 Me, s); 5.67 & 6.58 (d, 10 Hz)]-DMP

Table 3 (continued)

Compound	Aromatic protons			Protons of groups on the aromatic ring and other protons
	H-4	H-5	Others	
25. Heptazolicine	8.3		6.7–7.2(m)	[3.03, 1.98(t); 1.45(2 Me, s)]-DMHP
26. Dihydroxygirinimbine (220)	7.72(br, s)	7.96(8 Hz)	7.24, 7.50, 7.06(dt 8 Hz each)	2.28(ArM, d, 1 Hz); [1.24 & 1.52, (2 Me, s); 3.79 & 4.98 (d, 8 Hz each]-DMHP
27. Murrayaquinone B	—	7.98(d, 9 Hz)	7.02(d, 9 Hz, H-6)	3.91(ArOM, s); 2.13(ArM, d, 1.5 Hz); 6.42(q, 1.5 Hz, H-2); [1.74 & 1.85(2 Me, s); 3.59 (d, 7 Hz); 5.23 (br, t, 7 Hz)]-DMA
28. Pyrayaquinone A (25)	—	7.79(s)	6.83(s)	[1.48(2 Me, s); 6.48 & 5.72(d, 10 Hz each)]-DMP; 2.16(d, 1.5 Hz, vinyl methyl); 6.46(q, 1.5 Hz, H-2)
29. Pyrayaquinone B (26)	—	7.94(d, 9 Hz)	6.86(d, 9 Hz)	[1.48 (2 Me, s); 6.60 & 5.70 (d, each 10 Hz)]-DMP; 2.14(vinyl methyl); 6.44(q, 1.5 Hz, H-2)
30. Mahanimbinol	7.50(s)	7.83(m)	6.76–7.23 (m; H-6, H-7, H-8)	2.30(ArM, s), [1.58, 1.73 & 1.83 (3 Me, s), 2.05 (4H allylic methylene); 3.43(Bz-CH_2, d, 7 Hz); 5.0 & 5.39 (olefinic 7.0, 5.7 Hz each)]-geranyl chain
31. Mahanimboline	7.8(s)	8.1(d)	7.25–7.5(m)	[2.6 (br, 4 deshielded, s, methylene-H); 5.7 & 7.0 (olefinic 2H, d); 6.0–5.8(m, 2 terminal methylene-H and C$\underline{H}$OH proton]-C_{10} unit.
32. Murrayazolinol (227)	7.5(s)	7.95(d, 7 & 1 Hz)	7.15–7.4(m)	2.34(ArM, s), 1.93 & 1.5(2 Me, s), 1.3(Me, s), 3.33(bz-CH_2, br), 3.8(carbinyl-H, d, 7.2 Hz)
33. Isomurrayazoline	—	—	7.95–7.13(m)	2.35(ArM, s); 3.3(m, Bz-CH_2); 1.47(6 methylene protons, and cyclohexane ring protons
34. Murrayaquinone C (17)	—	7.98(d, 9 Hz)	6.4(q, 2 Hz, H-2); 7.01 (d, 9 Hz, H-6)	2.13(ArM, d, 2 Hz); 3.91(ArOM, s); [1.56, 1.61 and 1.85(3 Me, s); 3.58(d, 7 Hz); 5.26(t, 7 Hz); 2.05(4H, s), 5.07(1H, m)]-geranyl chain

Compound			
35. Murrayaquinone D (**16**)	—	7.88 (d, 9 Hz)	6.85 (d, 9 Hz, H-6)	2.12(ArM, d, 2 Hz), 5.52(ArOH, s); [1.58, 1.64 and 1.85(3 Me, s), 2.07(4H, s), 3.51(d, 7 Hz), 5.0(1H, m), 5.32(t, 7 Hz)]-geranyl chain; 6.41(q, 2 Hz, H-2)
36. Pyrayaquinone C (**27**)	—	7.94 (d, 8.7 Hz)	6.86 (d, 8.7 Hz, H-6)	2.15 (allylmethyl, d, 1.7 Hz); 6.45 (olefinic, d, 1.7 Hz, H-2); [1.44(Me, s), 5.68 & 6.63 (each d, 9.8 Hz); 5.09 (t, 1.3 Hz), 2.12 (m, 2H), 1.76 (1H, m); 1.65 and 1.56 (2 Me, s)]-C_{10} unit
37. (±) Murrafoline (**33**)	—	—	6.70–7.90 (m, 13 H)	2.28 & 2.37 (2 ArM, s); 1.54 (vinyl methyl, s) [1.40(Me, s); 1.44 (2 Me, s)]-3-oxygen linked methyls: 4.58 (1H, t, 8 Hz); 3.24(1H, m)
38. Murrafoline B (**19**)	7.79 (s); 7.44 (s)	7.85 (br, d, 8 Hz); 7.98 (d, 8 Hz)	7.06, 7.11(dt, 1 & 8 Hz); 6.93(br, d, 8 Hz); 7.44 (br, s,); 7.24 (t, 8 Hz)	2.48 & 2.49(2 ArM, s); 3.87(ArOM, s); [4.69 (bz-CH, dd, 7 & 11 Hz); 1.46 & 1.56(2 Me, s); 2.30(dd, 7 & 11 Hz); 2.38 (t, 11 Hz)]-DMHP
39. Murrafoline C (**34**)	7.79 (s); 7.85 (s)	7.83 & 7.85 (d, 8 Hz each)	7.04 & 7.09 (dt, 1 & 8 Hz each); 7.24 (t, 8 Hz); 7.43 & 6.95 (br, d, 8 Hz each)	2.24 (ArM, s); 2.47(ArM, s), [1.54 & 1.42(2 Me, s); 2.27(dd, 7 & 11 Hz); 2.31(t, 11 Hz); 4.63 (dd, 7 & 11Hz)]-DMHP; [6.07(br, s); 5.56(d, 10 Hz); 1.38 & 1.41(2Me s)]-DMP
40. Murrafoline D (**157**)	7.73 (s); 7.43 (s)	7.85(dd, 1 & 8 Hz); 8.04 (br, s)	6.73 (s); 7.24(br, d, 8 Hz)	2.41 & 2.50 (2 ArM, s); 3.99 (ArOM, s); 1.40 & 1.51(2 Me, s), 2.14 (t, 12 Hz); 2.27 (dd, 7 & 12 Hz)
41. Murrafoline E (**229**)	7.58 (d, 7.8 Hz); 7.34 (s)	—	6.90(s, H-2), 6.92(d, 10 Hz); 7.58 (s); 6.99, 7.35, 7.47 and 8.07 (d, 7.8 Hz each); 7.13 & 7.20(t, 7.8 Hz each)	2.51 (ArM, s); 3.94 (ArOM, s); 6.01 (2H, Bz-CH_2-bonded to N); [1.41 (2Me s); 5.79 & 6.92 (each d, 10 Hz)]-DMP
42. Murrafoline F (**228**)	7.74 (s); 7.89 (s)	—	8.06; 7.54; 7.37; 7.99; 7.52; 7.44; 7.17(d, 1H, 7.8 H each); 7.34(d, 8.4 Hz, H-2′)	2.38 (ArM, s); 3.93 (ArOM, s); 4.14(ArOM, s); 4.43 (2H, s)

Table 3 (continued)

Compound	Aromatic protons			Protons of groups on the aromatic ring and other protons
	H-4	H-5	Others	
43. Bismurrayafoline A (35)	7.32(s); 7.36(s)	7.72; 7.85(d, 7 Hz)	6.62 (H-2 & H-2'); 7.0–7.6 (m)	2.46(2 ArM, s); 3.71, 3.82 (2 ArOM, s); 5.83 (Bz-CH$_2$ bonded to N)
44. Bismurrayafoline B (36)	7.73 (2H, s)	7.69 (2H, d, 8 Hz)	6.79 (d, 2H, 8 Hz)	2.46 (2 ArM, s); 3.86 (2 ArOM); 6.78 (2H, s, OH); [1.24(2Me, s); 1.28(2Me, s); 3.36(4H, d, 8 Hz); 5.01 (2H, m)]-2DMA
45. Bismurrayafolinol (37)	7.74(s)	7.92 & 8.07 (d, 1 Hz each)	6.78 (s, H-2); 7.00 (s, H-2'); 7.10–7.50 (m)	3.83, 3.97 (2 ArOM, s); 6.01 (2H, s, CH$_2$ bonded to N); 4.86(2H, s, bz-CH$_2$ bonded to oxygen)
46. Oxydimurrayafoline (38)	7.68 (d, 1 Hz each)	8.03 (dd, 7 & 1 Hz each)	7.46 (2H, d,); 7.40 & 7.22 (1H, d, 7 & 1 Hz each); 6.98 (d, 1 Hz, H-2)	4.76 (4H, s, 2 Bz-CH$_2$); 4.01 (2 ArOM, s)
47. Carbazomycin A (6)	—	8.25 (dd, 7 & 2 Hz)	7.13–7.42 (m)	2.40(2 ArM, s); 3.92 & 4.13 (2 ArOM, s)
48. Carbazomycin B (5)	—	8.23 (br, d, 6.8 Hz)	7.03–7.48 (m)	8.11 (ArOH); 2.39 & 2.35 (2 ArM, s); 3.75 (ArOM, s)
49. Carbazomycin C (39)	—	7.7 (d, 2 & 4 Hz)	7.31 (d, 8.8 Hz); 6.91 (dd, 8.8 & 2.4 Hz)	2.36 & 2.33(2 ArM, s); 3.84 & 3.74(2 ArOM, s); 8.06(ArOH, s)
50. Carbazomycin D (7)	—	7.71 (d, 2.4 Hz)	7.36 (d, 8.8 Hz); 6.97 (dd, 8.8 & 2.4 Hz)	2.40 & 2.32(2 ArM, s); 3.83, 3.87 & 4.06(3 ArOM, s each)
51. Carbazomycinal (11)	—	8.18 (br, d, 7.3 Hz)	7.21–7.46(m)	2.73 (ArM, s); 3.74(ArOM, s); 10.40(Ar-CHO, s); 9.51(ArOH, s)
52. 6-Methoxycarbazomycinal (24)	—	7.65 (br, d, 2.4 Hz)	6.97 (dd, 8.8 & 2.4 Hz); 7.60 (d, 8.8 Hz)	2.40(ArM, s); 3.72 & 3.83(ArOM, s); 10.31 (Ar-CHO, s); 9.56(Ar-OH, s)

No. Compound				Aromatic H	^{1}H signals (aliphatic, OCH$_3$, OH etc.)
53. Carbazomycin G (**40**)	—	—	8.05 (m)	7.21–7.50 (m)	1.60 (CH$_3$, s); 2.01 (ArM, s); 3.71 (ArOM, s)
54. Carbazomycin H (**230**)	—	—	7.66 (d, 2.4 Hz)	7.39 (d, 8.8 Hz); 6.85 (dd, 8.8 & 2.4 Hz)	1.66 (CH$_3$, s); 2.05 (ArM, s); 3.76 & 3.84 (2 ArOM, s)
55. Kinamycin A (**193**)	—	—	—	—	1.58 (s, tertiary CH$_3$); 2.1–2.2 (3, O-COCH$_3$ gr); 4.66 (dd, 1.8 & 8.5 Hz); 5.25 (d, 1.8 Hz, OH); 5.84 (d, 8.5 Hz) & 6.53 (s)
56. Kinamycin B (**195**)	—	—	—	—	1.54 (s, tertiary CH$_3$); 2.3 (s, O-COCH$_3$)
57. Kinamycin C (**42**)	—	—	—	7.6, 7.13 (dd, 8.0, 2.0 Hz each); 7.5 (t, 8.0 Hz)	1.3 (s, tertiary CH$_3$); 2.57 (s, OH); 2.0–2.3 (s, 3-0-COCH$_3$); 5.4 (s, H adjacent to O-COCH$_3$); 5.6 & 6.2 (d, 7.2 Hz, H adjacent to O-COCH$_3$); 12.0 (s, phenolic OH bonded)
58. Kinamycin D (**194**)	—	—	—	7.58 (t, 7.8 Hz); 7.71 (dd, 1.1 & 7.3 Hz); 7.22 (dd, 1.1 & 7.9 Hz)	1.22 (s, tertiary CH$_3$); 2.95 (br, OH); 2.27 (s, O-COCH$_3$); 5.59 (d, 8.1 Hz); 5.43 (s), 2.19 (s, O-COCH$_3$); 4.79 (d, 8.1 Hz); 5.50 (br, s, OH); 12.4 (s, ArOH bonded).
59. Kinamycin E (**240**)	—	—	—	7.22 (dd, 1.1 & 7.7 Hz); 7.71 (dd, 1.1 & 7.1 Hz); 9.58 (t, 7.9 Hz)	1.48 (s, tertiary CH$_3$); 2.8 (br, OH); 4.24 (d, 7.9 Hz); 3.0 (br); 5.53 (s); 2.62 (s, O-COCH$_3$) 5.71 (d, 1.3 Hz); 4.61 (dd, 1.3 b 7.9 Hz); 12.17 (s, ArOH bonded)
60. Kinamycin F (**233**)	—	—	—	6.95–7.65 (m, 3H)	3.8 (d, 3.3 Hz); 4.63 (d, 3.3 Hz); 1.3 (s, tertiary CH$_3$); 4.65 (s, H-1'); 3.75 (1H, d, 3.4 Hz); 4.55 (d, 3.4 Hz)
61. Prekinamycin (**190**)	—	—	—	7.04 (dd, 0.9 & 7.2 Hz); 7.17 (dd, 7.0 & 7.2 Hz); 7.24 (dd, 0.9 & 7.0 Hz); 6.60 & 6.69 (d, 9.1 Hz each)	2.39 (s, ArCH$_3$); 11.60 (s, ArOH); 12.32 (s, phenolic OH bonded)
62. Ketoanydrokinamycin (**186**)	—	—	—	7.27 (d); 7.63 (t, 7.7 Hz); 7.76 (d, 7.6 Hz)	1.64 (s, tertiary CH$_3$); 3.89 (d, 2.7 Hz); 5.94 (d, 1.6 Hz); 5.34 (dd, 1.7 & 2.9 Hz); 12.06 (s, Phenolic OH bonded)

Table 3 (continued)

Compound	Aromatic protons			Protons of groups on the aromatic ring and other protons
	H-4	H-5	Others	
63. Tubingensin A (20)	7.92 (s)	7.98 (br, d, 7.8 Hz)	7.18 (dd, 3.9, 7.6 & 7.8 Hz); 7.11 (s); 7.34 (m)	2.99 (ddd, 7.3, 12.9, 17.6 Hz); 2.88 (br, d, 6.6, 17.6 Hz); 1.52 & 2.01 (m); 1.74 (m); 1.7 & 1.17 (m); 1.66 & 2.05 (m); 4.99 (br, s, OH); 1.71 & 2.08 (m); 1.76 & 2.06 (m, 2H); 5.03 (dd, 6.6 & 5.9 Hz, vinylic); 1.43 (br, s, CH_3); 1.21 (s, CH_3)
64. Tubingensin B (21)	8.05 (br, s)	7.98 (br, 7.8 Hz)	7.15 (ddd, 1.5, 7.3 & 7.8 Hz); 7.32 (ddd, 1.1, 6.6, 7.3 Hz); 7.3 (dd, 1.5 & 6.6 Hz); 7.32 (br, s)	1.82 (ddd, 4.4, 12, 14.1 Hz); 1.4 (m); 1.24 (m); 0.17 (ddd, 5.0, 13.4, 14.1 Hz); 1.22 (m); 1.31 & 1.72 (m, each); 2.11 (ddd, 3.2, 2.4, 10.7 Hz); 2.68 (m); 4.36 (br, t); 1.76 (m); 2.59 (br, dd, 3.6, 10.3 Hz); 1.44 (m); 1.66 (m); 2.41 (qq, 6.6, 6.8 Hz); 1.00 (d, 6.6 Hz); 1.09 (d, 6.8 Hz); 0.70 (d, 6.8 Hz); 1.15 (s, CH_3)
65. Hyellazole (91)	7.70 (s)	8.08 (d, 8 Hz)	7.6–7.35 (m); 7.28 (dd, 7.5, 1 Hz); 7.11 (dd, 7 & 1 Hz, H-6 & H-7)	3.99 (ArOM, s); 2.14 (s, ArM)
66. 6-chlorohellazole (41)	7.75 (br, s)	8.13 (d, 2 Hz)	7.6–7.35 (m); 7.27 (dd, 8.5, 2 Hz, H-7)	3.99 (ArOM, s); 2.14 (s, ArM)

(c) ^{13}C-*NMR Spectra*: Chemical shifts of C-1 to C-8 in the ^{13}C NMR spectra of carbazoles (Table 4) occur in the range δ 110–δ 126.5 [*107*]. Signals of C-1 and C-8 (δ 110.5) which are *ortho* to the imino nitrogen exhibit identical diamagnetic shifts (δ 18.02) compared with the shift of CH in benzene (δ 128.7) and are closer to the value of C-7 of indole (**22**) (*118*). Shifts of C-2 and C-7 of carbazole comparable with the shift of C-2 of indole. The signals of Carbons 8a and 9a at the ring junction appear at δ 140.2, further downfield than C-9 of indole. This may be due to greater delocalisation in the carbazole system. Substituents on the carbazole ring induce chemical shifts similar to those found in other aromatic systems; thus due to shift effect of the neighbouring hydroxy or methoxy group, signals of C-1 in 1-hydroxy-3-formylcarbazole and of C-6 in carbazomycin C are found at δ 146 and δ 154 respectively. Carbons in *ortho*- and *para*-positions experience induced shifts similar to those in benzene. C-1 in 2-oxygenated carbazoles experiences shift effects due both to oxygen and nitrogen atom; thus in alkaloids like mukonal and 2-hydroxy-3-methylcarbazole the signal of C-1 appears at δ 96.3 and δ 96.0 respectively (*11*).

Studies on several N-substituted carbazoles showed that the transmission of electronic effects through the nitrogen in carbazoles as evidenced by chemical shifts is close to that of aniline (**23**) (*40*).

(**22**)   (**23**)   (**24**)

Shifts of carbon signals in nitro- and bromocarbazoles have been reported (*116*). In 6-Methoxycarbazomycinal (**24**) selective proton decoupling (LSPD) by irradiation of H-2, H-5 caused the collapse of the C-6, C-12; C-7, C-13; C-5 and C-13 signals (*74*).

Modern NMR techniques like 2D NMR (*75*), COSY, HETCOSY, and NOESY have been utilized in obtaining structural information for various alkaloids. ^{13}C-NMR data of some alkaloids other than those reported previously (*11*) are given in Table 4.

4. Mass Spectra

The characteristic carbazolopyrilium ion at m/z 248 observed (*18*) in the mass spectra of pyranocarbazoles like girinimbine (**3**) and mahanimbine (**10**) is paralleled in the mass spectra of carbazoloquinones like

Table 4. ^{13}C-NMR Data of Carbazole and Some of Its Derivatives (δ-values)

Name of the compounds	C-1	C-2	C-3	C-4	C-4a	C-4b	C-5	C-6	C-7	C-8	C-8a	C-9a	C-10	C-11	C-12	C-13	C-14
1. Carbazole (1)	110.5 (s)	125.8	119.4	120.3	119.2	119.2	120.3	119.4	125.8	110.5	140.5	140.5	—	—	—	—	—
2. Carbazomycin C (39)	109.8 (s)	127.9 (s)	139.0 (s)	143.6 (s)	110.8 (s)	124.7 (s)	106.2 (d)	154.0 (s)	113.9 (d)	111.5 (d)	135.6 (s)	138.9 (s)	13.4 (q)	12.8 (q)	61.3 (q)	—	56.0 (q)
3. Carbazomycin D (7)	114.5 (s)	129.0 (s)	144.5 (s)	146.6 (s)	115.0 (s)	123.8 (s)	105.9 (d)	154.4 (s)	114.5 (d)	112.0 (d)	135.9 (s)	138.7 (s)	13.7 (q)	12.7 (q)	61.0 (q)	60.6 (q)	56.0 (q)
4. Carbazomycinal (11) (Carbazomycin E)	111.2 (s)	134.8 (s)	139.6 (s)	154.4 (s)	111.9 (s)	122.6 (s)	123.1 (d)	120.5 (d)	121.6 (s)	111.9 (d)	140.7 (s)	139.6 (s)	189.4 (s)	11.4 (q)	61.3 (q)	—	—
5. 6-Methoxy-carbazomycinal (Carbazomycinal F) (24)	110.3 (s)	135.0 (s)	139.0 (s)	151.9 (s)	111.7 (s)	123.3 (s)	106.5 (d)	154.8 (d)	115.4 (s)	112.4 (d)	140.1 (s)	139.9 (s)	189.3 (s)	11.3 (q)	61.3 (q)	—	56.1 (q)
6. Carbazomycin G (40)	67.3 (s)	154.3 (s)	147.6 (s)	177.5 (d)	108.4 (s)	123.8 (s)	121.5 (d)	120.5 (d)	122.9 (d)	112.0 (s)	136.4 (s)	140.8 (s)	27.9 (q)	10.1 (q)	59.2 (q)	—	—
7. Carbazomycin H (230)	67.2 (s)	154.4 (s)	147.5 (s)	177.4 (s)	108.3 (s)	124.4 (s)	102.3 (d)	155.0 (s)	112.4 (d)	112.7 (d)	131.1 (s)	140.6 (s)	27.9 (q)	10.1 (q)	59.1 (q)	—	55.2 (q)
8. Tubingensin A (only carbons of tricyclic carbazole system given) (20)	110.65	135.08	132.43	118.45	121.32	123.78	119.77	119.13	110.41	139.99	137.84						
9. Tubingensin B (only carbons of tricyclic carbazole system given) (21)	107.39 (d)	142.03 (d)	134.80 (d)	116.83 (d)	119.80 (s)	123.65 (s)	119.70 (d)	119.19 (d)	125.08 (d)	110.48 (s)	139.62 (s)	137.95 (s)					
10. O-Demethylmurrayanine (217)	146.0 (s)	108.4 (d)	118.0 (s)	127.3 (d)	109.5 (s)	116.0 (s)	121.0 (d)	120.1 (d)	126.0 (d)	112.7 (d)	145.2 (s)	131.3 (s)	—	—	194.1 (d)	—	—
11. N-methoxy-3-formylcarbazole (219)	108.0 (s)	108.5 (d)	119.7 (s)	120.9 (d)	121.2 (s)	123.9 (s)	127.2 (d)	127.8 (d)	129.5 (d)	—	137.5 (s)	140.2 (s)	—	—	191.7 (d)	—	—
12. Murrayafoline A (12)	145.2 (s)	107.7 (d)	125.4 (d)	128.0 (s)	110.9 (s)	123.4 (s)	110.9 (d)	120.3 (d)	124.3 (d)	112.5 (d)	139.5 (s)	129.2 (s)	55.1 (q)	—	21.8 (q)	—	—
13. Murrayaquinone A (8)	180.4 (s)	136.0 (s)	131.6 (d)	183.4 (s)	124.1 (s)	116.2 (s)	126.1 (d)	123.7 (d)	122.2 (d)	113.7 (d)	137.8 (s)	148.3 (s)	—	—	16.0 (q)	—	—

C_{10}–C_{14} represent carbons substituted on aromatic rings.

(25–27) by a base peak at m/z 278 representing the quinonoid carbazolopyrilium ion (28, 29).

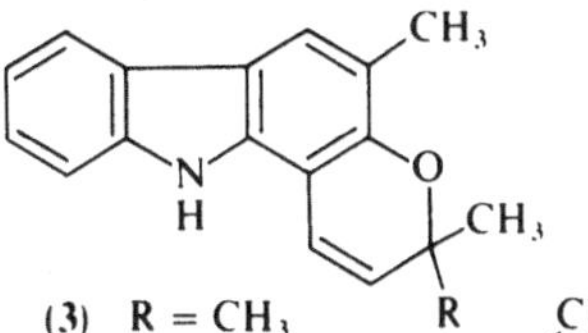

(3) R = CH₃
(10) R = CH₂CH₂CH = C(CH₃)(CH₃)

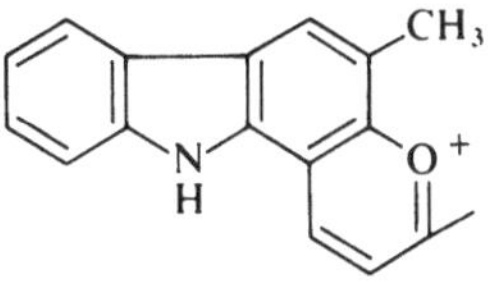

m.z 248

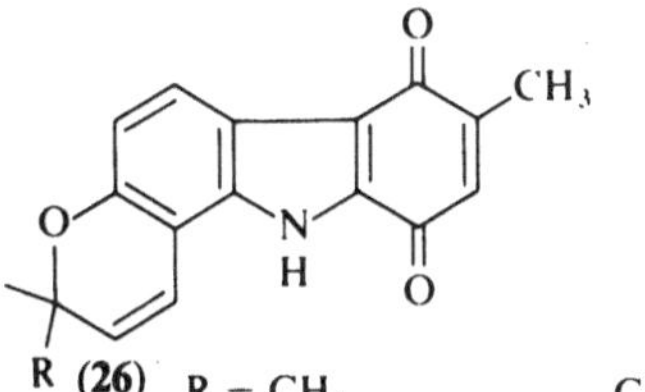

(26) R = CH₃
(27) R = CH₂CH₂CH = C(CH₃)(CH₃)

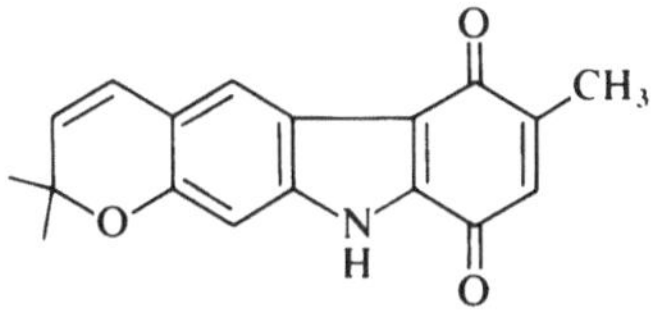

(25)

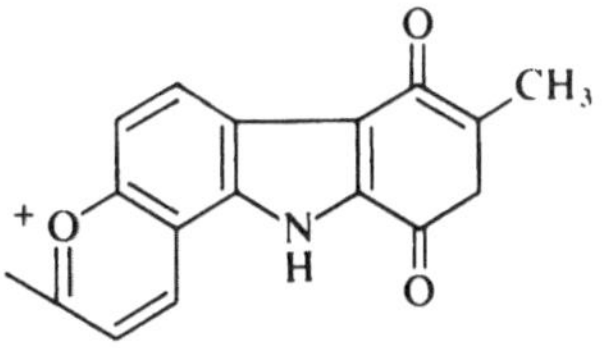

(28)

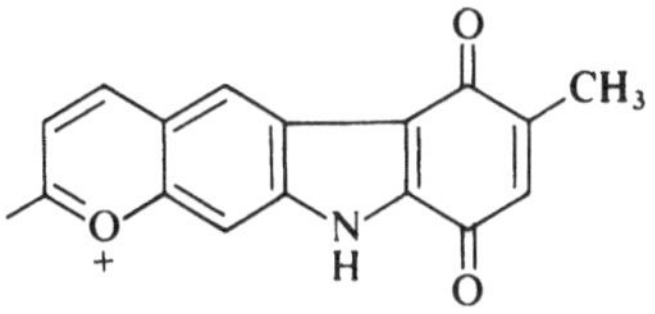

(29)

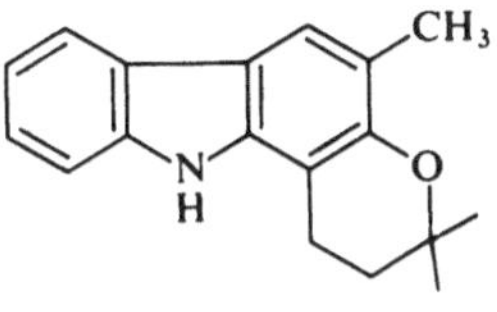

(31)

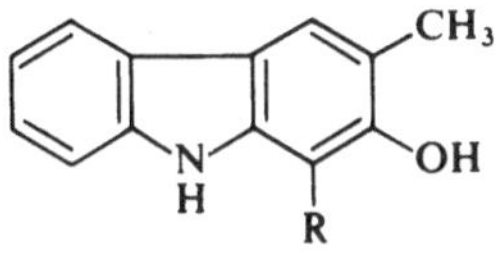

(30) R = Geranyl chain

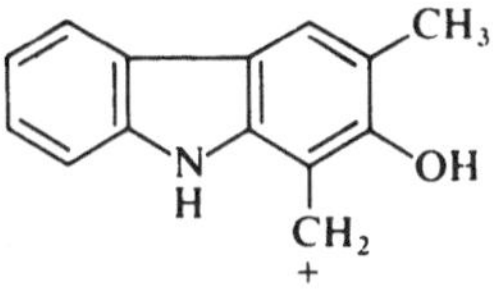

(32) m/z 210

The peak at m/z 210 in the mass spectra of mahanimbiol (30) and alkaloids like dihydrogirinimbine (31) could be rationalised in terms of ionic species (32) observed in the mass spectra of many saturated pyranocarbazoles (*18*).

The appearance of appropriate molecular diion (M^{++}) peaks (Table 5) in the mass spectra of several alkaloids as shown below has been utilized for ascertaining the binary nature of alkaloids.

Table 5. *Molecular Di(+)ion Peaks of Some Bisalkaloids*

Sl.	Alkaloid	M^+ (m/z)	M^{++} (m/z)	Ref.
I	(±) Murrafoline (33)	594.3218	297	(*81*)
II	Murrafoline B (19)	474.2325	237	(*44*)
III	Murrafoline C (34)	526.2618	256	(*44*)
IV	Bismurrayafoline A (35)	420.0000	210	(*45*)
V	Bismurrayafoline B (36)	588.29829	294	(*45*)
VI	Bismurrayafolinol (37)	436.1789	210	(*58*)
VII	Oxydimurrayafoline (38)	436.1809	211	(*58*)

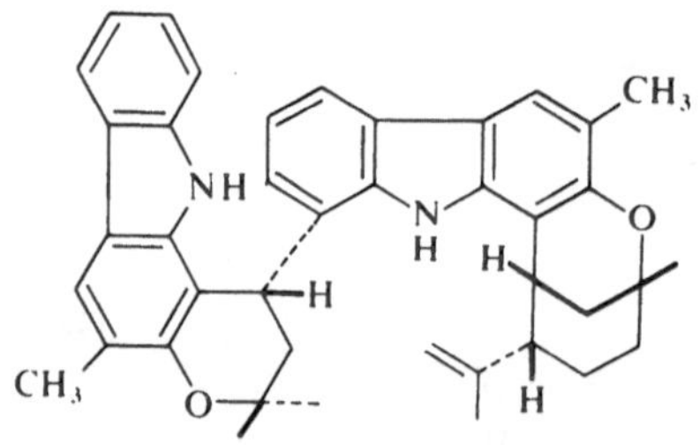

(33)

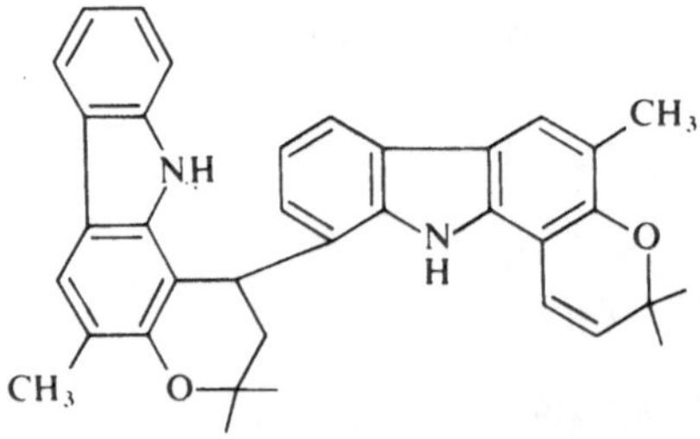

(34)

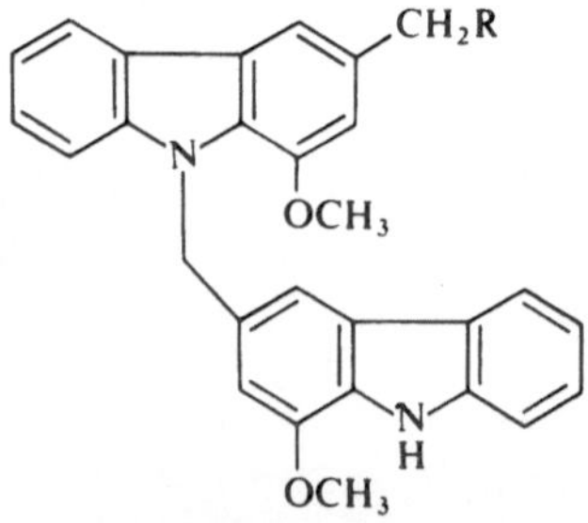

(35) R = H

(37) R = OH

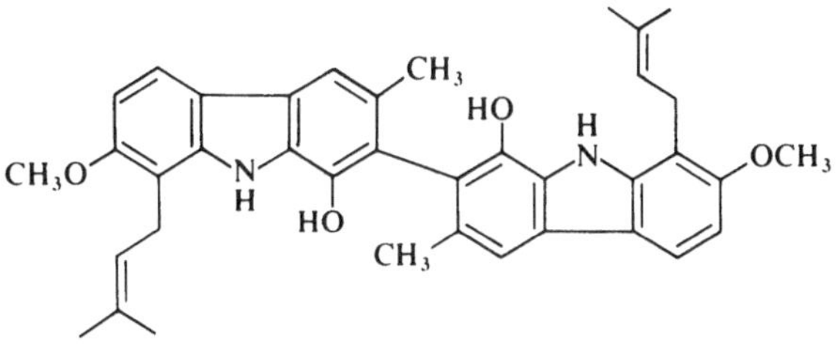

(36)

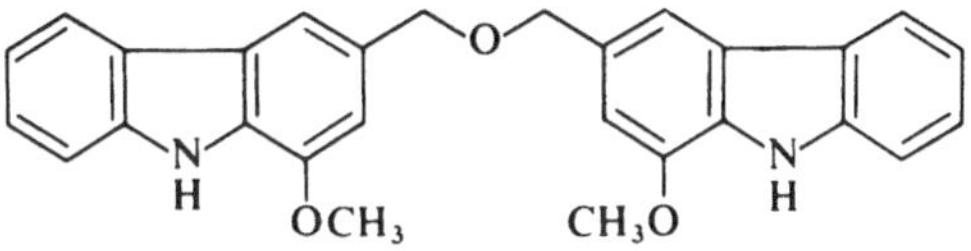

(38)

5. X-ray Crystallographic Methods

X-ray crystallographic methods have been used to confirm the structures of carbazomycin-B (5), -C (39), -G (40) (66) and chlorohyellazole (41) (11). The N-CN function in the pyrrole fragment and the overall structure comprising the benzo-b-carbazole system of the kinamycin group of alkaloids have been deduced by X-ray diffraction (47) although chemical methods were also used for assignment of the nitrile function of kinamycin C (42) to the carbazolic nitrogen.

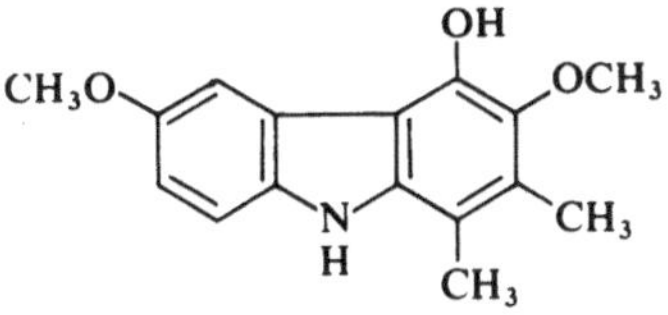

(39)

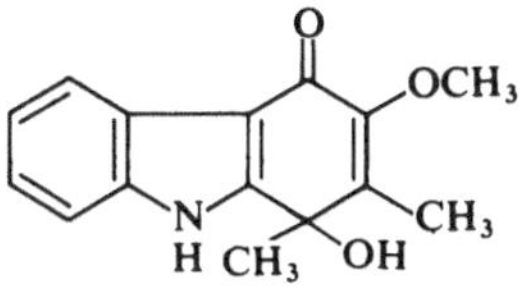

(40)

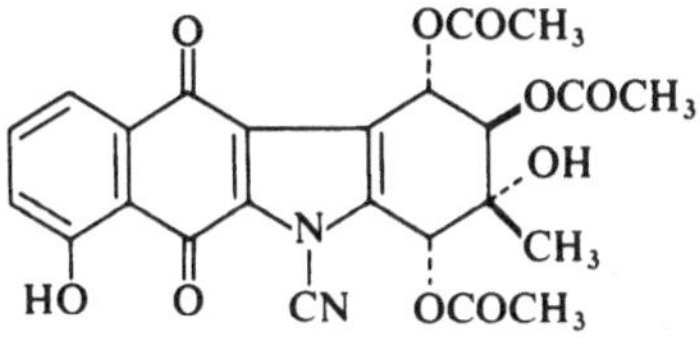

(42)

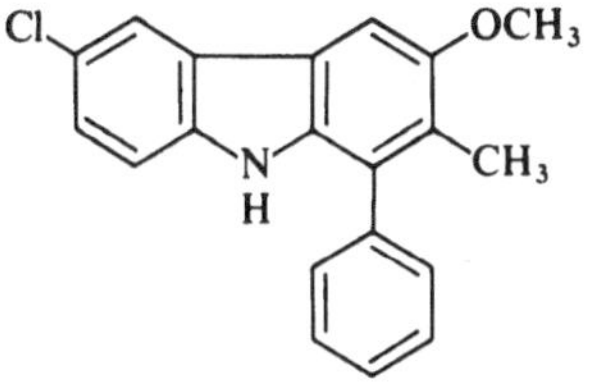

(41)

B. Degradative Reactions

Due to the availability of well-established physical data and newer physical methods, applications of degradation methods for structure determination of carbazole alkaloids during the period under review have been fewer. However zinc dust distillation and some reactions mentioned previously (*18*) have been used for characterizing some of the alkaloids (*11, 107, 24*). Reduction of phenol tosylates has been utilized to confirm presence of a phenolic group in some cases (*11, 107*). Hydrolytic degradation of kinamycin C (**42**) to establish the presence of a nitrile group on the carbazole nitrogen, is interesting as X-ray crystallography did not permit a decision between presence of a nitrile or an isonitrile function (*89*).

C. Synthesis of Carbazoles

Previous reviews relating to the synthesis of carbazoles and newer methods of synthesis have been cited by JOULE in an updated account (*63*). Methods related to recent syntheses of carbazole alkaloids are briefly enumerated here.

1. Synthesis of Tricyclic Alkaloids

(a) From Arylamines

(i) *Electrophilic aromatic substitution of arylamines.* The method of electrophilic aromatic substitution of appropriately substituted aryl-amines at the *ortho* positions using an iron tricarbonyl hexadiene complex (**43**) previously used by BIRCH (*13*) has been utilized for the synthesis of carbazomycin A and B (*73*).

The electrophilic aromatic substitution of amine (**44**) by iron complexed cation (**43**) at room temperature provided (**48**) after three days while the complex (**45**) was obtained after two hours. Oxidative cyclisation of (**45**) as such was not facile. Acetylation to (**46**) followed by oxidation with MnO$_2$ gave O-acylcarbazomycin B (**47**) which on de-acetylation furnished carbazomycin B (**5**). However selective oxidation of (**48**) provided dihydrocarbazole-3-one (**49**). Demetallation with trimethylamine N-oxide afforded the 3-hydroxycarbazole (50) which on methylation afforded carbazomycin A (6).

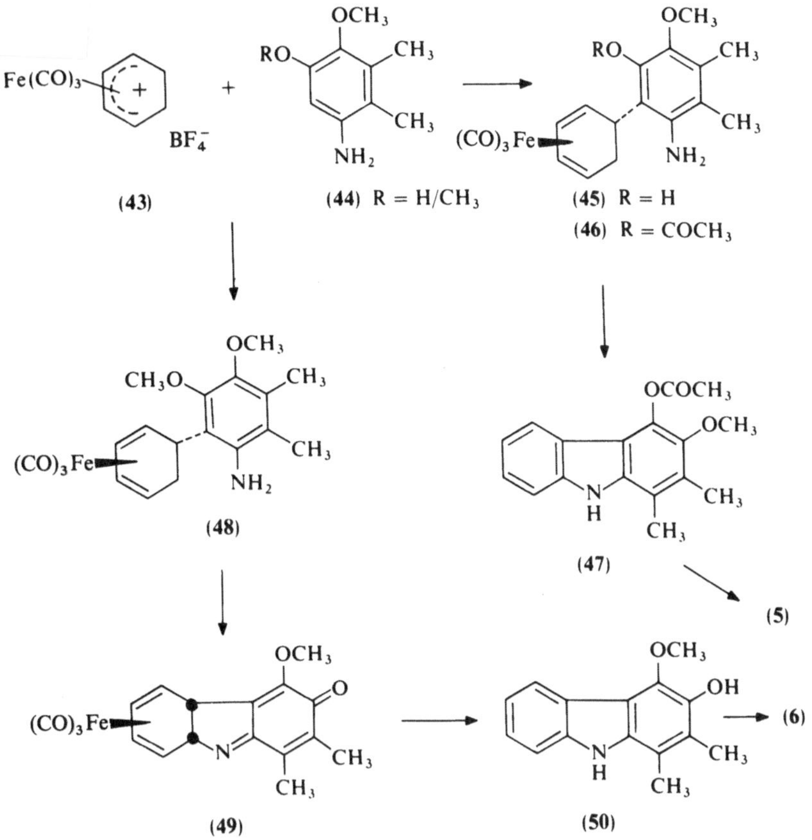

(ii) *Lewis-acid catalysed aliphatic diazocoupling*: Diazocoupling of cyclohexanones (51, 52) with the highly reactive electrophile benzene diazonium cations (53–54) under Lewis acid catalysis to provide the hydrazones (55, 56) required for the formation of the oxotetrahydro-carbazoles (57, 58) has been utilized by Roy *et al.* (*105*). This procedure eliminates the step for formylation of cyclohexanone necessary in the Japp–Klingemann reaction (*18*) to obtain the above oxotetrahydro-carbazole.

(iii) Iodine catalysed thermal cyclisation of anthranilic acid (60) has been found to yield carbazole (1) and diphenylamine (61) as major products along with other products (*12*).

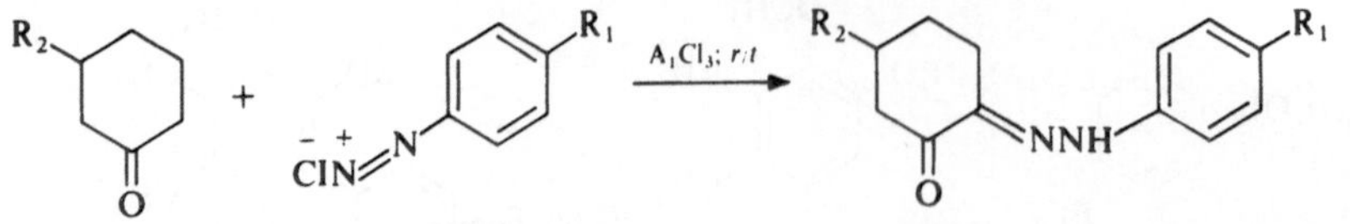

(51) R_2 = H

(52) R_2 = CH_3

(53) R_1 = H

(54) R_1 = OCH_3

(55) R_1 = R_2 = H

(56) R_1 = OCH_3; R_2 = CH_3

(1) R_1 = R_2 = H

(59) R_1 = OCH_3, R_2 = CH_3

(57) R_1 = R_2 = H

(58) R_1 = OCH_3, R_2 = CH_3

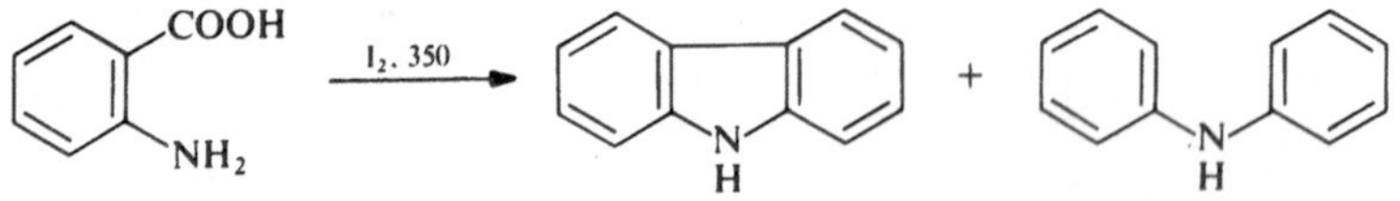

(60)

(1)

(61)

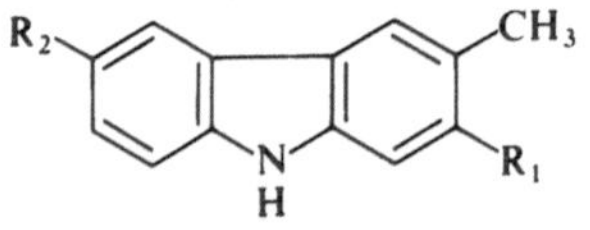

(62) R_1 = R_2 = H

(63) R_1 = R_2 = OCH_3

(b) From Diphenylamines

The cyclodehydrogenation of diphenylamine (61) to carbazole (1) has been accomplished with various catalysts like activated metal catalysts (62), and iodine (18). JACKSON and SASSE (62) used degassed Raney nickel, but obtained only poor yields while CHAKRABORTY and SEN GUPTA (113) using degassed Raney nickel in the presence of p-cymene in a sealed tube synthesised alkaloids like 3-methylcarbazole (62) glycozoline (59) and glycozolidine (63) from appropriate diphenylamines in better yields. However, palladium acetate is by far the best catalyst for this process. Recently, several alkaloids including murrayastine and murrayaline have been synthesised by this method (41).

Condensation of 1-bromo-2-methoxy-4-methyl benzene (64) with 5,6-dimethoxyaniline acetate (65) in pyridine in the presence of Cu and K_2CO_3 and subsequent hydrolysis with 20% KOH furnished the diphenylamine derivative (66) which on treatment with palladium acetate in DMF afforded murrayastine (13). Condensation of 1-bromo-3-methoxy-4-methylbenzene (67) with the dimethyl acetal of 6-formyl-5-methoxyaniline acetate (68) furnished the diphenylamine derivative (69) which on hydrolysis and on subsequent cyclisation furnished murrayaline (14).

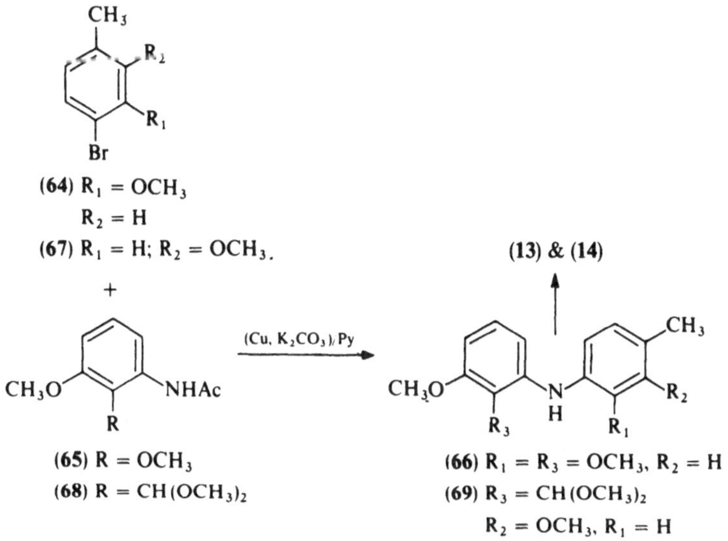

CH₃

(64) $R_1 = OCH_3$
       $R_2 = H$
(67) $R_1 = H$; $R_2 = OCH_3$.

(13) & (14)

+

CH_3O — NHAc

(65) $R = OCH_3$
(68) $R = CH(OCH_3)_2$

(Cu, K₂CO₃)/Py →

CH_3O — N — H

(66) $R_1 = R_3 = OCH_3$, $R_2 = H$
(69) $R_3 = CH(OCH_3)_2$
       $R_2 = OCH_3$, $R_1 = H$

Chart 1. Synthesis of murrayastine and murrayaline

(c) From Indoles

(x) From indole-2-carboxylates (79)

Indole-2-carboxylate (70) on Claisen condensation with butyrolactones (71, 72) gave the lactones (73, 74) which on subsequent hydrolysis and decarboxylation gave the alcohols (75, 76). On oxidation with pyridiniumchlorochromate (PCC) the alcohols gave the aldehydes (77, 78) which on cyclisation with BF_3-methanol gave 1-methoxycarbazole (74) derivatives. With 4-methylbutyrolactone (72) murrayafoline A (12) has been synthesised.

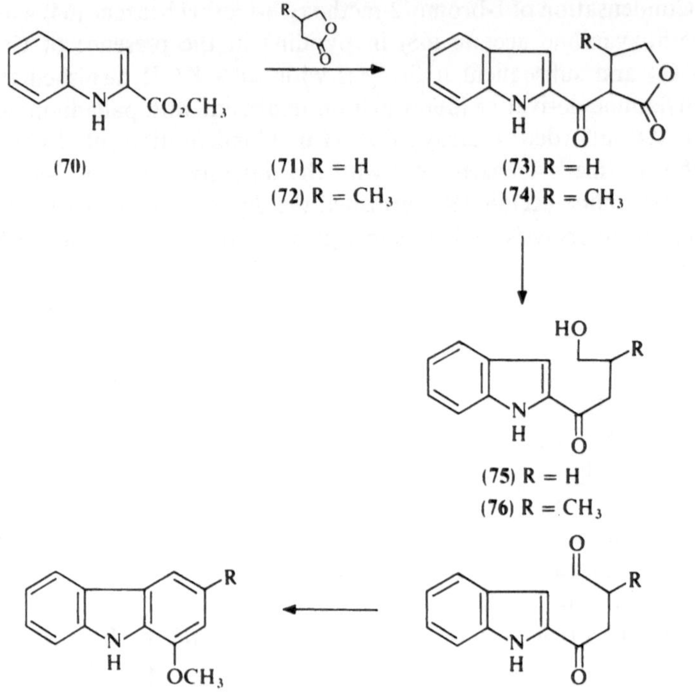

(70)

(71) R = H
(72) R = CH₃

(73) R = H
(74) R = CH₃

(75) R = H
(76) R = CH₃

(79) R = H: (12) R = CH₃

(77) R = H: (78) R = CH₃

(y) From 2-alkylated indoles

BERGMAN and CARLSON (8) reported the preparation of 2-hydroxy-3-methylcarbazole (83), the key compound for synthesis of pyranocarbazole alkaloids and using aldehyde (81) alkylate 2-methylindole (80) at the 3-position (82). Use of 2,3 unsaturated ketones (84) and 2-

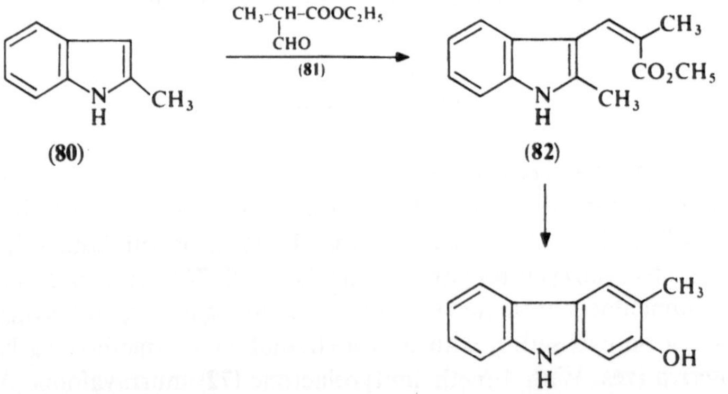

(80)

(82)

(83)

methylindoles (**80**) in the presence of Pd/C gave better yields of 2-methylcarbazole (**85**) and various related compounds (*7*). The use of molecular sieve in the reaction mixture improves the yield.

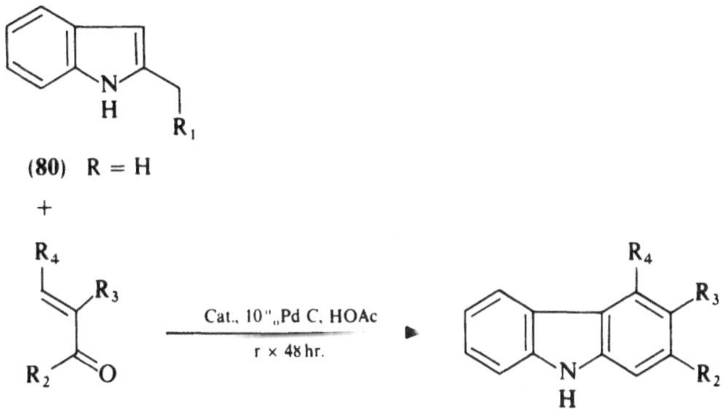

(**80**) R = H

+

Cat., 10 %, Pd C, HOAc
r × 48 hr.

(**84**) $R_2 = CH_3$
$R_3 = R_4 = H$

(**85**) $R_2 = CH_3$
$R_3 = R_4 = H$

(z) Cycloaddition reactions involving indoles

(i) *From 3-vinylindoles*: The occurrence of the 3-vinylindole system imbedded in carbazomycin B, girinimbine & pyridocarbazoles prompted investigations (*1*) on the synthesis of carbazoles using 3-vinylindoles (**86**). Hence (4 + 2)-cycloaddition of a dienophile to a vinylindole has been utilized to build tricyclic and polycyclic systems (*98*). The synthesis of 4-demethoxycarbazomycin A (*100*) and its regioisomer represents an interesting application of this method.

(4 + 2) cycloaddition of (**87**) and dimethyl acetylenedicarboxylate and dehydrogenation of the cycloadduct gave (**88**) which on hydrolysis and subsequent reduction furnished 4-demethoxycarbazomycin A (**89**).

The regioisomer 3-demethoxycarbazomycin A was prepared starting from indolyl(methoxy)methylcarbenium tetrafluoroborate (**91**) which was deprotonated *in situ* to provide a 3-vinylindole (**91a**). This on trapping with dimethyl acetylenedicarboxylate and subsequent dehydrogenation and reduction gave 3-demethoxycarbazomycin A (**90**) through (**92**).

(ii) *From 2-vinylindoles*. Use has also been made of 2-vinyl-indole (**93**) in spite of its lesser reactivity as compared with the 3-vinyl isomer. The syntheses of hyellazole (**94**) and chlorohyellazole (**41**) are good examples (*11*). BERGMAN and PELCMAN (*7*) synthesised 3-demethoxy-hyellazole (**95**) starting from substituted 2-vinylindole (**96**) through several steps (**97–100**). PINDUR as well as other authors have used 2-vinylindole for the synthesis of carbazoles.

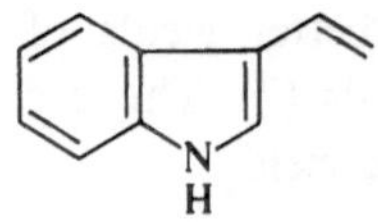

The structural aspects of 3-vinyl-1H-indoles for predicting the out-come of Diels-Alder reactions has been worked out and reviewed by PINDUR *et al.* (*101*, *97*).

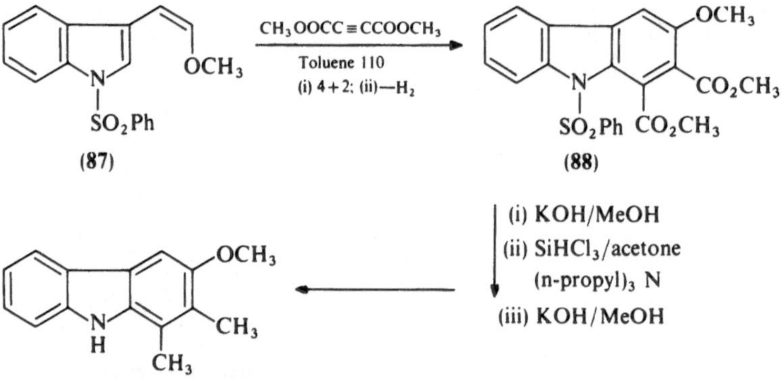

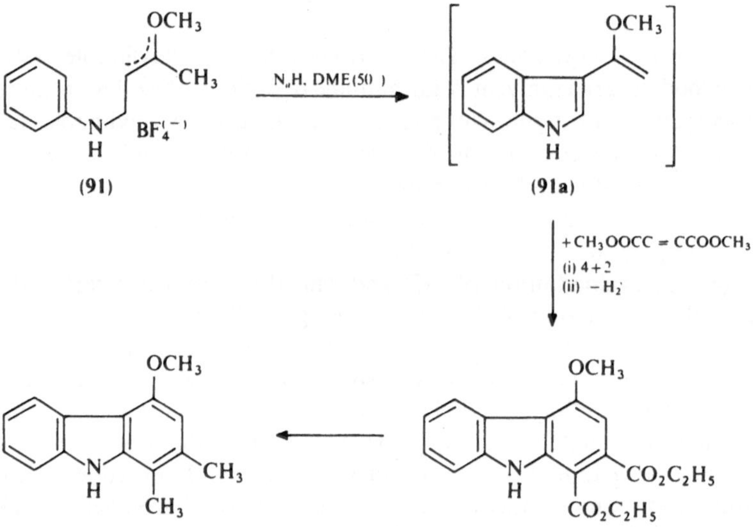

(iii) *From indole-2,3-quinodimethane and their analogues* (*99*). The potential of indole 2,3-quinodimethane (**101**) as a synthetic tool was first realised by MAGNUS and coworkers in their synthesis of complex indole

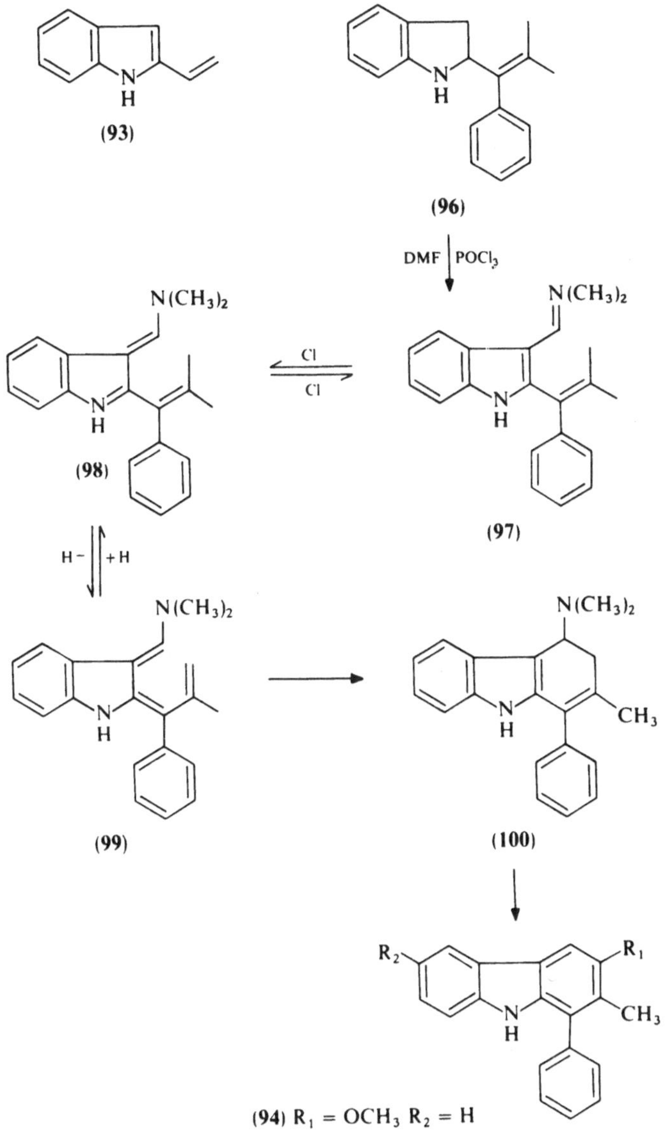

(93)

(96)

DMF | POCl₃

N(CH₃)₂

(98)

Cl
Cl

N(CH₃)₂

(97)

H⁻ ‖ +H

N(CH₃)₂

(99)

N(CH₃)₂

CH₃

(100)

R₂

R₁

CH₃

(94) R₁ = OCH₃ R₂ = H
(95) R₁ = R₂ = H; (41) R₁ = OCH₃, R₂ = Cl

alkaloids (48): This reactive species and its analogues have been used extensively for the synthesis of carbazoles. Pyrano-(3,4-b)-indole-3-one (102), an analogue of (101), has been utilized in the synthesis of carbazomycin B and heyllazole by MOODY and SHAH (84). Diels-Alder reaction of (102A) with 3-trimethyl silyl propynoate (103) gave carbazole

(**104**) which was reduced to 2-methylcarbazole derivative (**105**) with $LiAlH_4$. Mercurodesilylation of (**105**) gave (**106**) which on hydroboration and oxidation gave the phenol (**107**). The latter's methylation gave 4-deoxycarbazomycin B (**89**) identical with the substance prepared by PINDUR (*100*). Starting with (**102B**) and following the sequence so far described, hyellazole (**94**) was obtained. To obtain carbazomycin from 4-deoxycarbazomycin B the NH-group was protected with the t-butoxycarbonyl group (**108**). The latter on treatment with N-bromosuccinimide in acetonitrile, gave the corresponding 4-bromo derivatives (**109**). Sub-

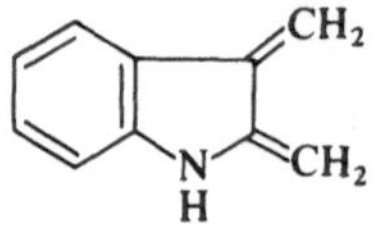
CH2
N
H
CH2

(**101**)

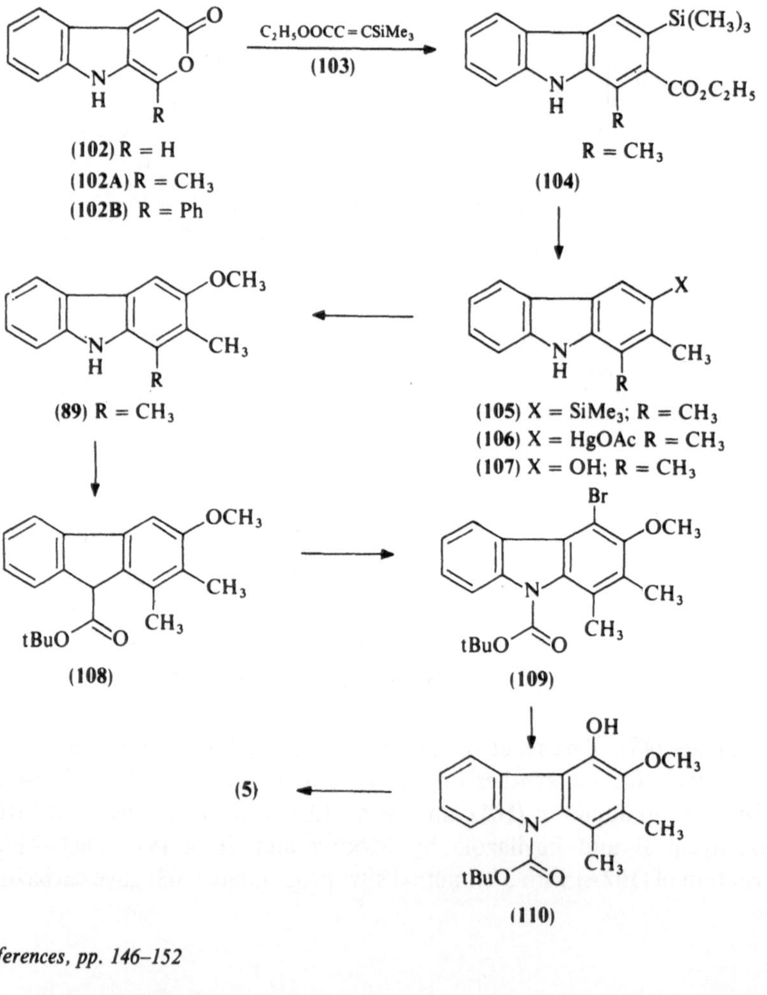
O
N
H
O
R
(102) R = H
(102A) R = CH3
(102B) R = Ph
C2H5OOCC≡CSiMe3
(103)
Si(CH3)3
CO2C2H5
N
H
R
R = CH3
(104)
OCH3
CH3
N
H
R
(89) R = CH3
X
CH3
N
H
R
(105) X = SiMe3; R = CH3
(106) X = HgOAc R = CH3
(107) X = OH; R = CH3
OCH3
CH3
CH3
tBuO
O
(108)
Br
OCH3
CH3
N
CH3
tBuO
O
(109)
OH
OCH3
CH3
N
CH3
tBuO
O
(110)
(5)

sequent treatment with t-butyl lithium in THF at $-78\,°C$ followed by reaction of the aryllithium with trimethyl borate and alkaline hydrogen peroxide work up provided the 4-hydroxycarbazole (110). Removal of the tertiary butoxycarbonyl group gave carbazomycin B (5).

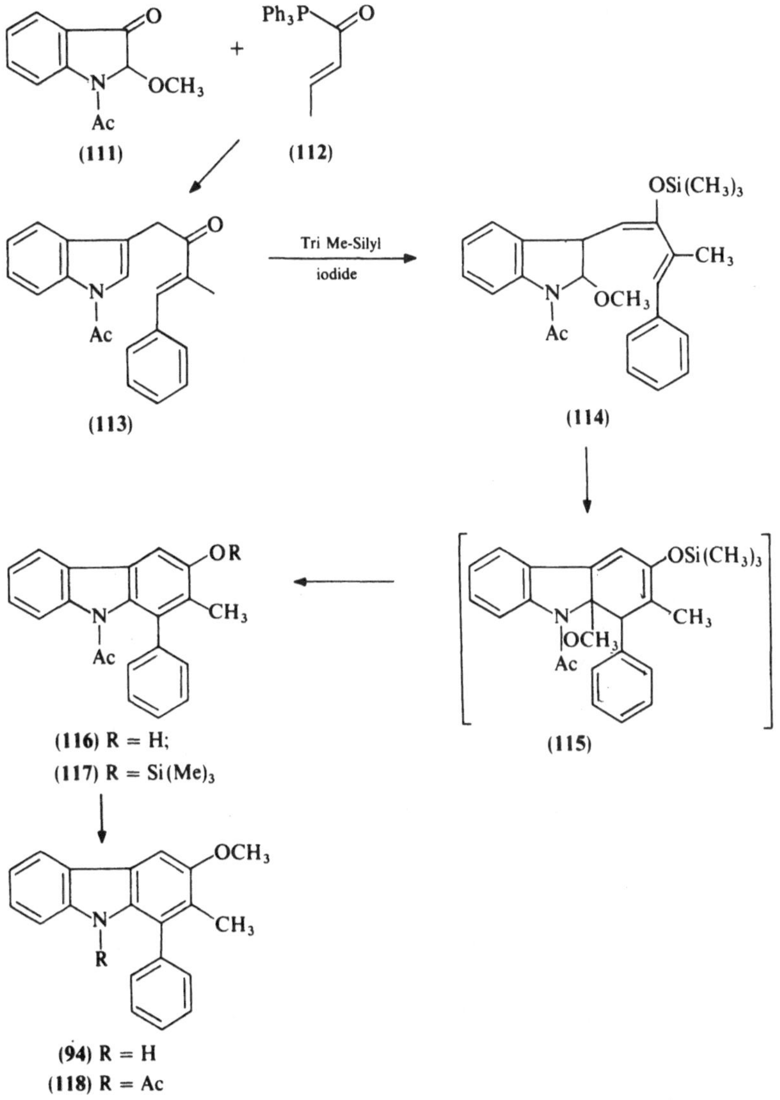

(iv) *Benzannelation of indoles* (70): A convenient synthesis of hyeallazole has been effected by benzannelation of an indole. 2-Methoxy-indoline-3-one (111) on Wittig Reaction with phosphonium ylide (112) gave (113) which on treatment with trimethyl-silyliodide in presence of hexamethyldisilazane (HMDS) provided the desired 3-buta-1,3-dienylindole (**114**) in 80% yield. Heating to (**114**) induced an electrocyclic reaction leading to (**115**) from which methanol was readily eliminated giving rise to the 3-hydroxycarbazole derivative (**116**) and its silyl ether (**117**). The silyl ether on treatment with *t*-butylammonium fluoride (TBAF) afforded (**116**). This on methylation (**118**) and deacetylation furnished hyellazole (**94**).

(d) Biomimetic hydroxylation of 3-methylcarbazole (*106*)

Synthesis of 2-hydroxy-3-methylcarbazole (**83**) has also been accomplished from 3-methylcarbazole (**62**) in the Udenfriend reaction (*122*), which is a prototype of biological hydroxylation with mixed function oxidases and monoxygenases.

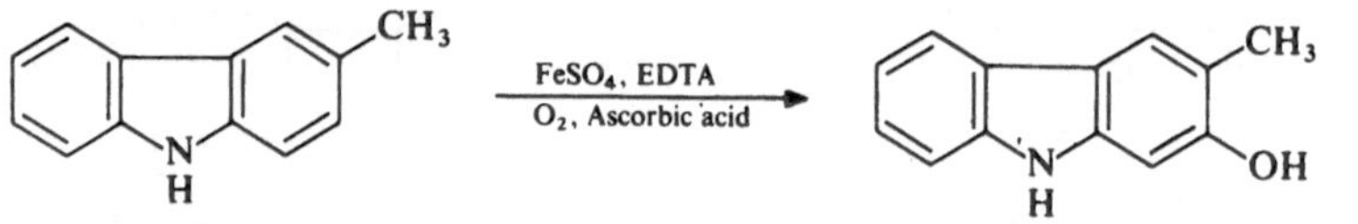

$$\xrightarrow[\text{O}_2,\ \text{Ascorbic acid}]{\text{FeSO}_4,\ \text{EDTA}}$$

(62)   (83)

2. *Synthesis of Tetracyclic Alkaloids*

(a) C_{18}-Alkaloids

(i) Diphenylamine route has been used for the synthesis of pyrano carbazoles like pyrayafoline and pyrayaquinones -A and -B.

For the synthesis of pyrayafoline (*41*), 1-bromo-3-methoxy-4-methylbenzene (**67**) was condensed with 2,3 [2',2' dimethyl-$\Delta^{3'}$-pyrano-]-aniline acetate (**119**) to furnish the diphenylamine derivative (**120**). Cyclisation with Pd(OAc)$_2$ in DMF then furnished pyrayafoline (**121**) along with its regioisomer.

For synthesis of pyrayaquinone A and pyrayaquinone $\bar{\text{B}}$ (*42*) 7-Amino-2,2-dimethyl-chromene (**122**) prepared from *m*-hydroxyacetamide (**123**) gave on condensation with 2-methylbenzoquinone (**124**) diphenylamine (**125**) which on cyclisation with palladium acetate afforded pyrayaquinone A (**25**). Similarly 5-amino-2,2-dimethylchromene (**126**) on condensation with 2-methylbenzoquinone (**124**) afforded the diphenylamine derivative (**127**) which on cyclisation by the above method furnished pyrayaquinone B (**26**).

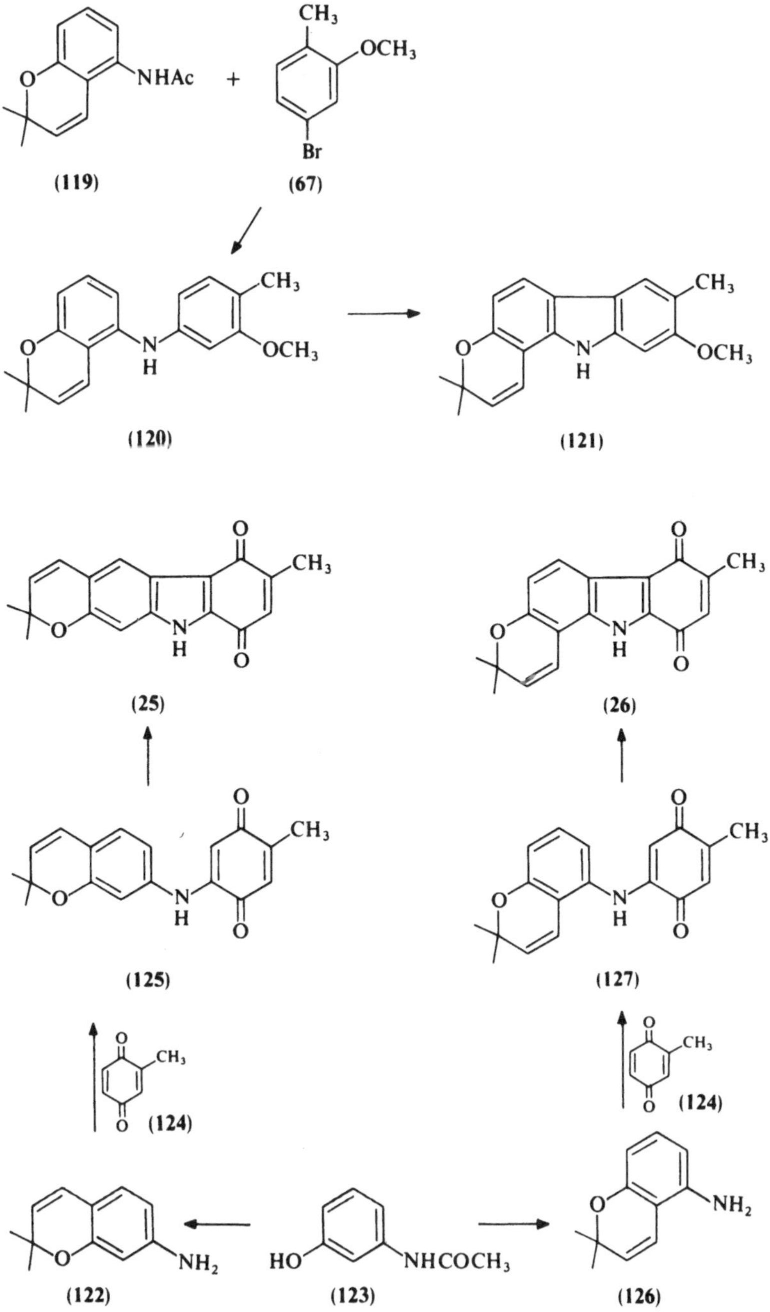

Chart 2. Synthesis of pyraquinone A and B

(ii) *Introduction of a C_5-unit with 3,3-dimethylacrylic acid (23)*

2-Hydroxycarbazolecarboxylic acid (**128**) on treatment with 3,3-dimethylacrylic acid at 145° in the presence of SbCl₃ furnished the indolochromanones (**129** & **130**). Both chromanones on reduction and on tosyl dehydrogenation furnished the linear pyranocarbazole (**131**) and its regioisomer nor-girinimbine (**132**), the linear chromanone being obtained in better yield.

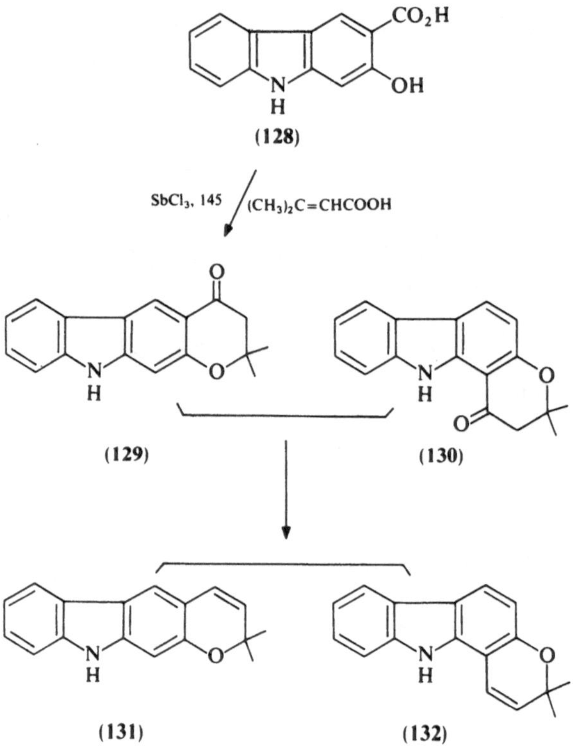

Chart 3. Synthesis of linear pyranocarbazole

On the other hand 1-hydroxy-3-methylcarbazole (**133**) reacted with 3,3-dimethylacrylic acid in the presence of zinc chloride and phosphorus oxychloride (*93*) to furnish indolochromanone (**134**) which on reduction and tosyl dehydrogenation furnished the girinimbine isomer (**135**).

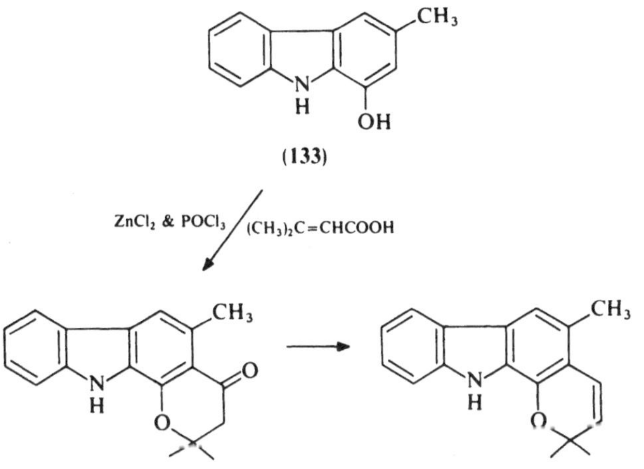

(133)

ZnCl$_2$ & POCl$_3$ / (CH$_3$)$_2$C=CHCOOH

(134) → (135)

(b) Benzo-b-Carbazoles and Carbazoloquinones

Methods for the synthesis of tetracyclic benzocarbazoles and benzo-b-carbazoloquinone have become relevant to alkaloidal synthesis due to the progress in the chemistry of the kinamycin group antibiotics.

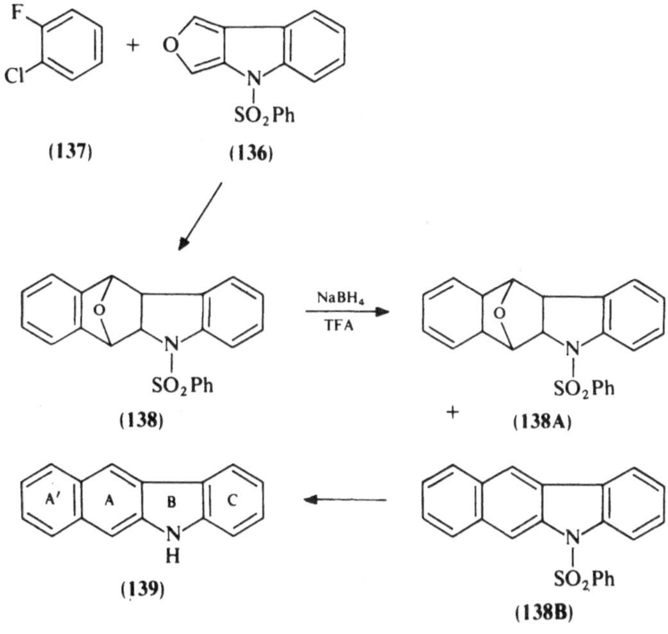

(137) + (136)

(138) → NaBH$_4$ / TFA → (138A)

+

(139) ← (138B)

Indole-2,3-quinodimethane methodology has been utilized in the synthesis of the above groups of carbazoles (*108*). The N-sulphophenyl indole-2,3-quinodimethane anologue (**136**) undergoes the Diels-Alder reaction with benzyne generated from (**137**). In three steps (**138–138B**) it furnishes the simple benzo-b-carbazole (**139**).

On the other hand a benzocarbazoloquinone (**141**) has recently been synthesised through oxidative cyclisation of a 2-benzoylindole containing benzylic alcohol group in the *ortho* position (**140**) with tetra-n-butyl-ammonium perruthenate (TBAP) (*51*).

(140)   (141)

(c) Tetracyclic C_{23}-Alkaloids and Their Pentacyclic
and Hexacyclic Cyclomers

Since the hexacyclic and pentacyclic bases synthesised so far are cyclomers of C_{23}-alkaloids of the mahanimbine group, their syntheses are discussed together with syntheses of tetracyclic alkaloids of C_{23} skeleton.

Syntheses of nor-mahanimbine and its regioisomers have been effected (*65*) by reacting 2-hydroxycarbazole (**142**) with citral (**148**) in boiling pyridine showing that the reaction is regioselective, not regiospecific.

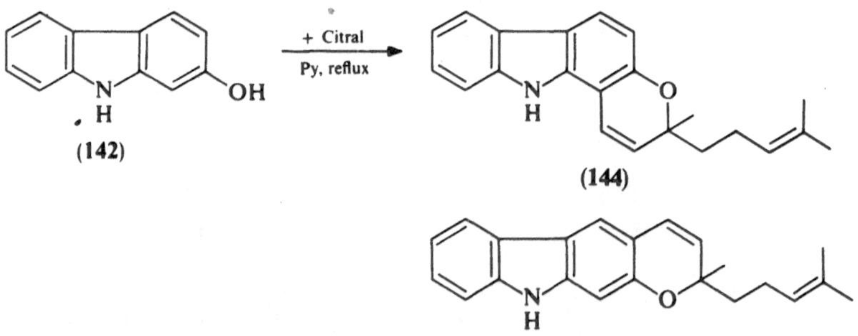

(142)   (144)

(143)

1-Hydroxycarbazole containing a methyl group at different positions of the carbazole nucleus produces different bases (*91*) when reacted with citral (**148**) in a mixture of acetic acid and butanol at 110°. A mixture of acetic acid and butanol has been found to be a better condensing agent for the citral condensation (*92*) than those reported previously. Compound (**145**) gave the mahanimbine isomer (**149**) while (**146**) and (**147**) gave the hexacyclic bases (**150**) and (**151**).

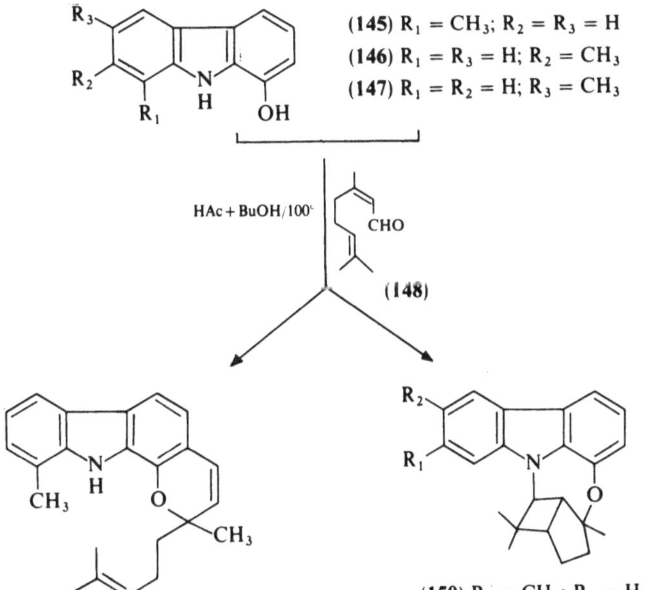

(**145**) $R_1 = CH_3; R_2 = R_3 = H$
(**146**) $R_1 = R_3 = H; R_2 = CH_3$
(**147**) $R_1 = R_2 = H; R_3 = CH_3$

(**150**) $R_1 = CH_3; R_2 = H$
(**151**) $R_1 = H; R_2 = CH_3$

Synthesis of bicyclomahanimbine (**152**) has been effected by CHAKRABORTY and DAS (*30*) who condensed photocitral A (**153**), obtained by irradiation of citral in sunlight or by refluxing it with pyridine, with 2-hydroxy-3-methylcarbazole (**83**).

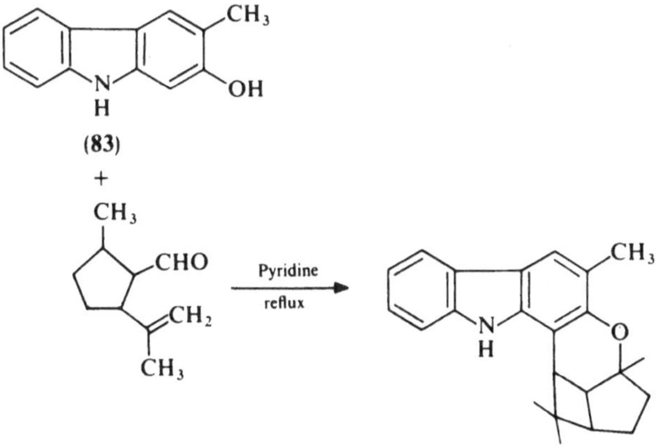

3. Synthesis of Biscarbazoles

(a) Formation of dimers with a C_{26}-skeleton:

(i) *Synthesis of Bismurrayafolinol Acetate (58)*
Murrayanine (2) on reduction with sodium borohydride gave 3-hydroxymethyl-1-methoxycarbazole (154) which on treatment with acetic anhydride furnished bismurrayafolinol acetate (155).

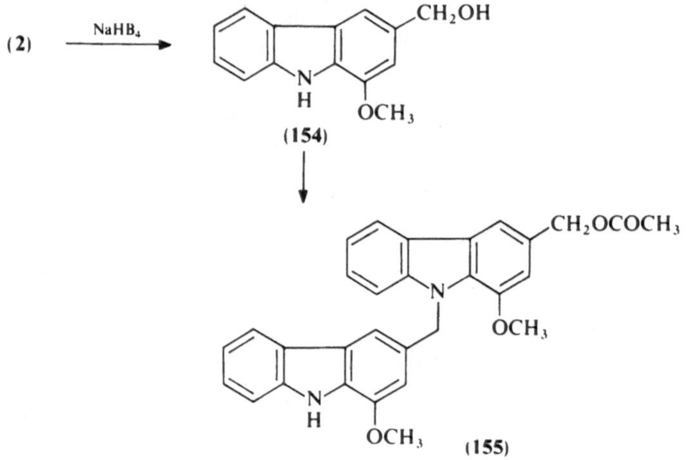

(ii) *Biomimetic dimerisation*: 3-Methylcarbazole (62) on Udenfriend reaction (122) furnished the biscarbazole (156) in poor yield (106).

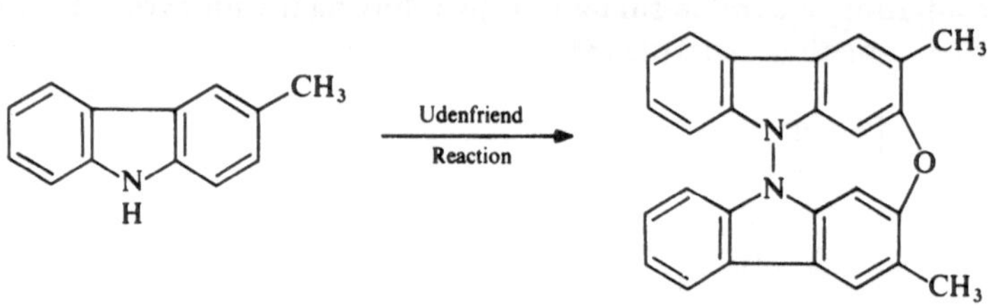

(b) Formation of dimers with C_{31}- and C_{36}-skeletons

(i) *Nafion 117 Catalysed Dimerisation.*
On refluxing murrayafoline A (12) and girinimbine (3) in aqueous methanol solution for 48 hours in the presence of Nafion 117, FURUKAWA *et al.* obtained three dimeric products (157, 157A, 158) (44).

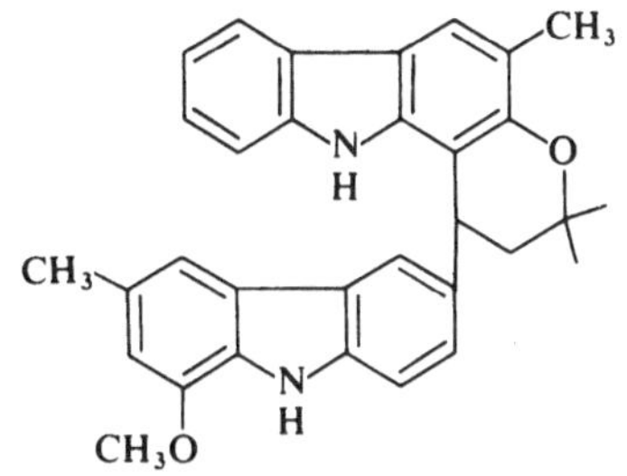

(157)

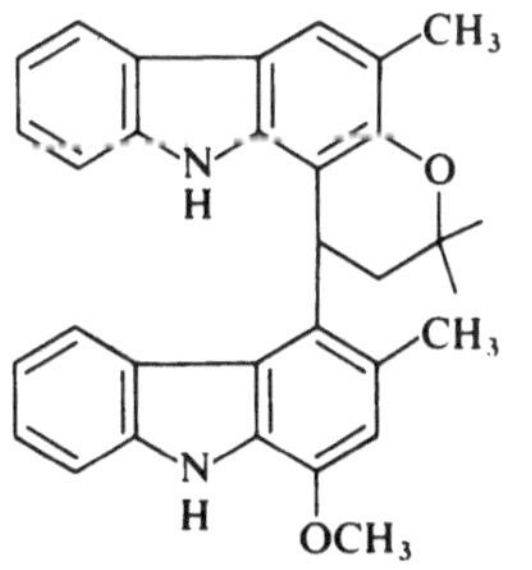

(157A) 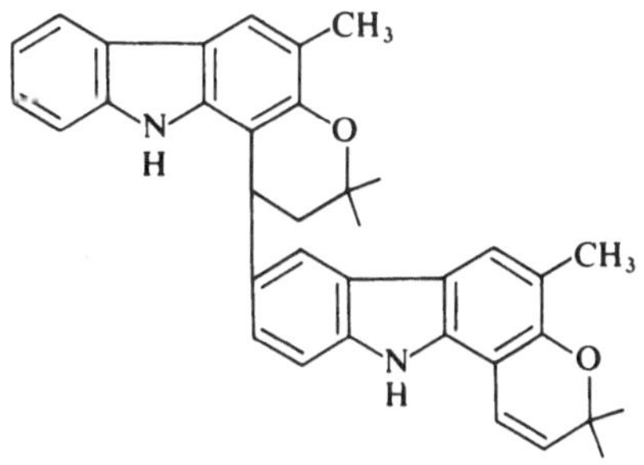   (158)

(ii) BF₃-Etherate Catalysed Dimerisation

CHAKRABARTI and CHAKRABORTY (16) have reported the BF₃-Etherate catalysed synthesis of biscarbazoles (158, 162, 163) starting with girinimbine (3) and its derivatives (159–160). Mechanistically the dimer formation has been envisaged as proceeding through the adduct (161).

4. Transformations and Photochemical Reactions

(a) Transformations

(i) Thermal Transformation of Mahanimbine

Mahanimbine (10) on heating at 200°C in a sealed tube for a short time is transformed to dl-mahanimbine first and then to murrayazoline (164) and murrayazolidine (165). On further heating for 3 hrs more only dl-murrayazolidine is obtained (87).

On passing mahanimbine (10) over a column of Dowex 50 × 8 (H⁺) resin dl-murrayazolidine (165) is obtained (5).

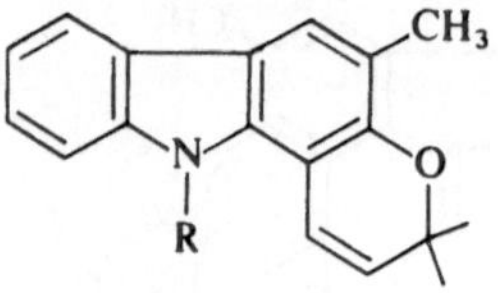

(3)   R = H
(159)   R = CH$_3$
(160)   R = C$_2$H$_5$

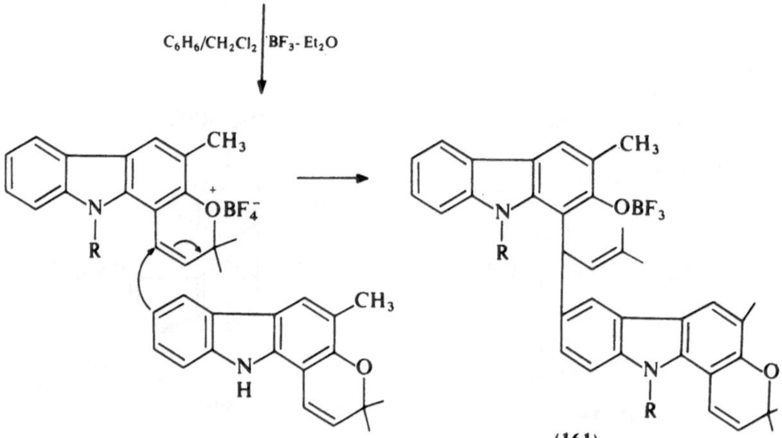

C_6H_6/CH_2Cl_2 | BF_3-Et_2O

(161)

10% NaHCO$_3$

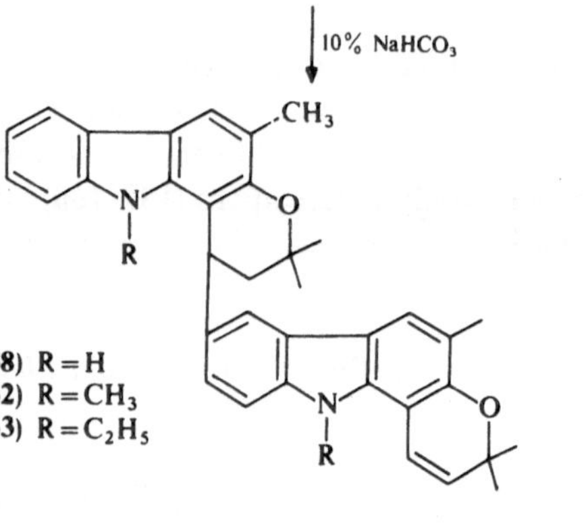

(158)   R = H
(162)   R = CH$_3$
(163)   R = C$_2$H$_5$

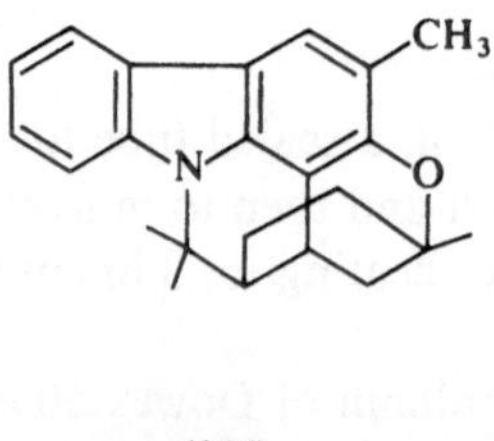

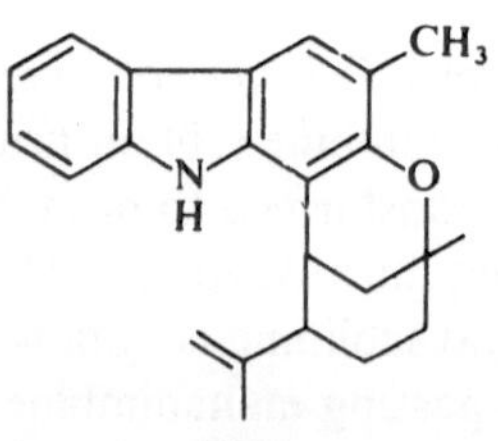

(164)

(165)

(ii) Racemisation and Inversion of Mahanimbine

Mahanimbine and normahanimbine (**144**) racemise thermally (*5*) at 90°C in isooctane solution in the dark. The racemisation has been attributed to facile transformation of the chromenes to enones (**166**) and (**167**) and subsequent ring closure. Mahanimbine on the other hand on standing in an ethanolic solution to six days undergoes optical inversion (*20*) from + 45° to −24.8. The inversion does not take place in CHCl$_3$ solution but is catalysed by adding a drop of concentrated HCl. The inversion can also be explained in terms of an enone–chromene trans-formation and subsequent ring closure. In attempts to isolate l-mahan-imbine after removing the solvent only dl-mahanimbine was obtained. In this connection the interesting photochemical transformation (*90*) of 2,2-dimethylchromene (**168**) in methanolic solution to (**169**) may be cited. Solvent addition to the intermediate enone species gives (**170**) which on heating regenerates the original chromene.

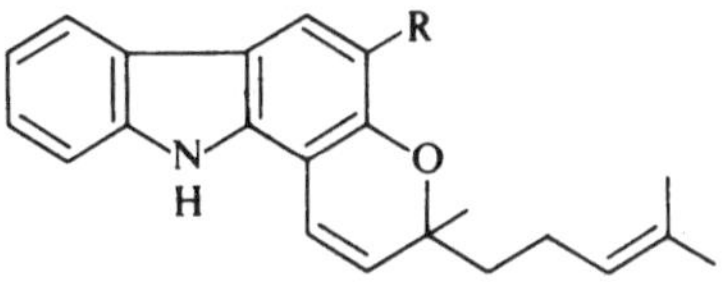

(**144**) R = H
(**10**) R = CH$_3$

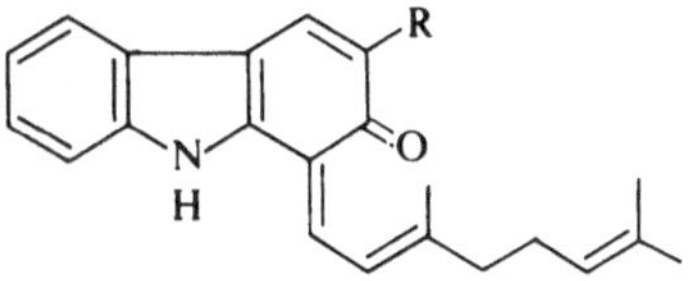

(**166**) R = H
(**167**) R = CH$_3$

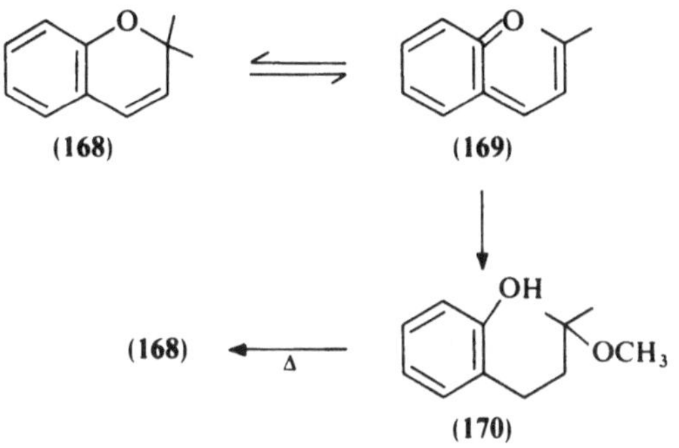

(**168**)

(**169**)

(**168**) ← Δ

(**170**)

(b) Photochemical Reactions

(i) Dimerisation of Vinylcarbazole

Photochemical syntheses of carbazoles and carbazole alkaloids from diphenylamine and other substrates have been discussed previously by

114   D. P. CHAKRABORTY and S. ROY

CHAKRABORTY (*18*) as well as by JOULE (*63*). Recently DE MAYO et al. (*3*) showed that vinyl carbazole (**171**) on photochemical reaction in CdS dispersed medium undergoes dimerisation leading to a cyclobutone (**172**) to which two carbazole units are attached.

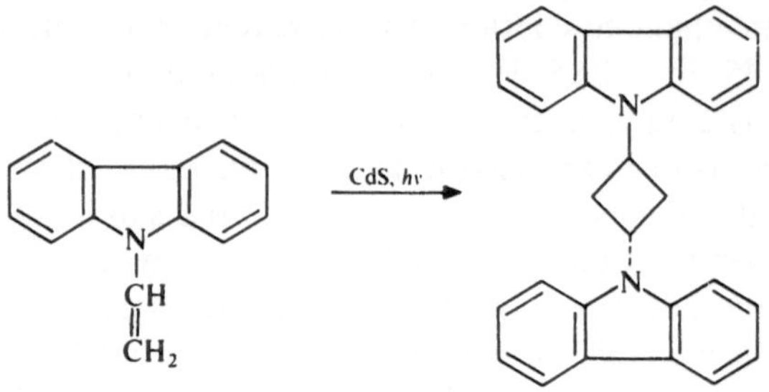

(171)   (172)

(ii) *Photo-Fries Rearrangement of N-Substituted Carbazoles*

SHIZUKE et al. (*115*) have shown that N-acetylcarbazole (**173**) in cyclohexane solution on irradiation with ultraviolet light undergoes a photo-Fries rearrangement to yield 1- and 3-acylcarbazole. They used ferric oxalate as an actinometer in their experiments. On the other hand, CHAKRABARTI and CHAKRABORTY (*17*) showed that N-sulphonyl carbazole (**174**) underwent smooth photo-Fries rearrangement on irradiation in benzene solution without any catalyst. Both groups invoked the solvent cage radical (**176**) as an intermediate through which the transformations took place. To explain the formation of carbazole (**1**) during this reaction,

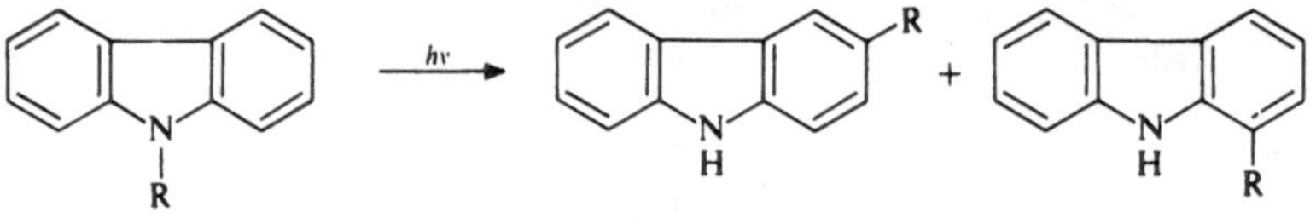

(173) R = COCH₃   (173–175a)   (173–175b)
(174) R = SO₂Ph   (173a, 173b) R = COCH₃
(175) R = COC₆H₅   (174a, 174b) R = SO₂Ph
                      (175a, 175b) R = COPh

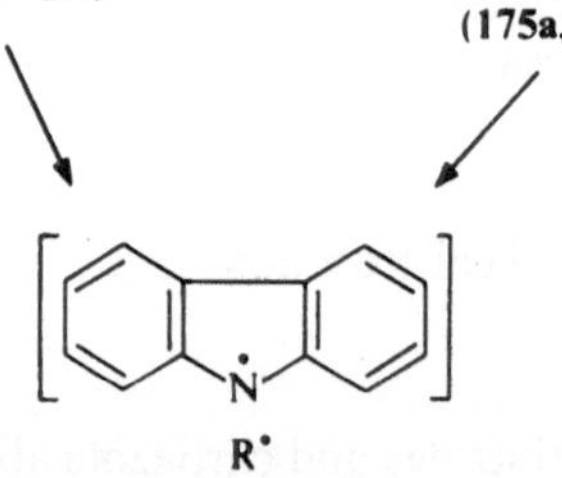

(176)

CHAKRABARTI and CHAKRABORTY suggested that the carbazole radical leaks from the solvent cage and abstracts hydrogen from other molecules to form (1). GHOSH *et al.* (*49*) on the other hand have shown that N-benzoyl carbazole (175) undergoes photo-Fries rearrangement only in the presence of iodine in methanolic solution through a separate mechanism.

(iii) *Ring Expansion of Pyranocarbazoles* (*15*)

The photolytic transformation of some pyranocarbazoles (177a–177d) and (3, 10) have been shown to yield a seven membered 2,5-dihydrooxepinocarbazole ring system (179a–179d), (3b, 10b). The ring expansion is assumed to proceed *via* an O-quinoellide intermediate of general structure (178); a subsequent 1,4 shift and recyclization forms the corresponding seven membered ring system. The sulphonyl derivative (177e) did not undergo any change showing the probable participation of the nitrogen lone pair in the overall rearrangement.

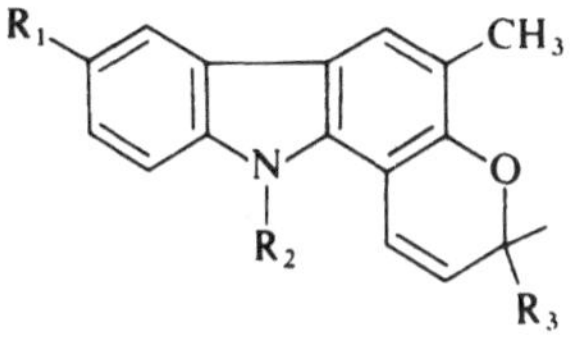

(177a–177e), 3, 10

(177a) $R_1 = OCH_3$; $R_2 = H$; $R_3 = CH_3$
(177b) $R_1 = H$; $R_2 = CH_2CH = CH_2$; $R_3 = CH_3$
(177c) $R_1 = H$; $R_2 = R_3 = CH_3$
(177d) $R_1 = H$; $R_2 = C_2H_5$; $R_3 = CH_3$
(177e) $R_1 = H$; $R_2 = SO_2CH_3$; $R_3 = CH_3$
(3) $R_1 = H$; $R_2 = H$; $R_3 = CH_3$
(10) $R_1 = R_2 = H$; $R_3 = CH_2CH_2CH = C(CH_3)_2$

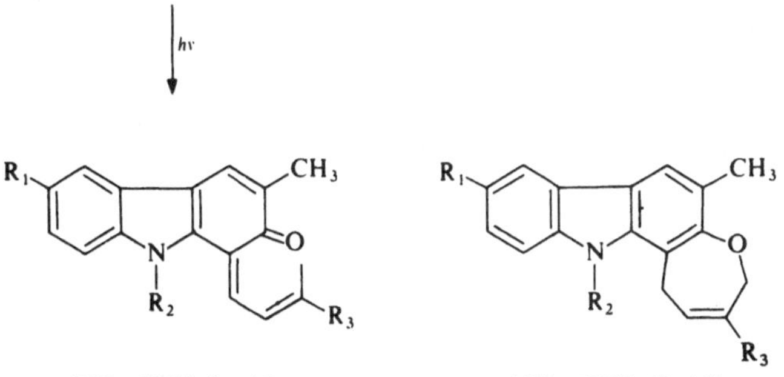

hv

(178a–178d), 3a, 10a   (179a–179d), 3b, 10b

(c) Formation of Bicyclodihydrooxepinocarbazole from
Dihydrooxepinocarbazoles

Some of the 2,5 dihydrooxepinocarbazoles of type (179) undergo
further transformations which leads to bicyclooxepino derivatives of type
(180). Formation of the cyclopropane ring may take place through a di-Π
methane rearrangement.

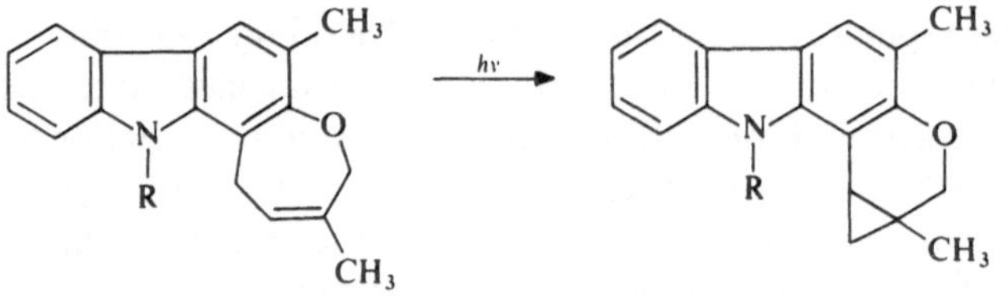

179(b) R = CH₃ – CH = CH₂
179(c) R = CH₃
179(d) R = C₂H₅

(180a) R = CH₃ – CH = CH₂
(180b) R = CH₃
(180c) R = C₂H₅

(d) Photochemical Synthesis of Benzocarbazoles

GRELMANN and SCHMITT (52) have shown that photochemical synthesis of benzocarbazole can be accomplished starting from N-methyl-2-
anilinonaphthalene (181). This reaction has close similarity to the ring
closure reaction of N-methyl diphenylamine (18). It has been proposed
that after intersystem crossing from the first excited singlet of (181) to its
excited triplet, ring closure occurs to the singlet ground state of the
zwitterionic intermediate (182). Formation of benzocarbazole (185) then
takes place through a complicated decay process whereby a small
amount is converted to dihydrocarbazole (184). The first step in the
rearrangement is a sigmatropic 1,4 shift in analogy with the rearrangement of enamines, to give dihydrocarbazole (183) which readily re-
arranges to the stable form (184). However the mechanism by which (184)
is converted to (185) is not very clear. The reaction has been shown to be
regioselective, the selectivity probably being governed by the electron
density distribution in the excited triplet state. Benzocarbazole (185A)
was not obtained as a reaction product. Incidentally the kinamycin
group of antibiotic alkaloids is built on a benzo-b-carbazole (185A)
skeleton. Improved syntheses of substituted carbazoles and benzocarbazoles via lithiation of dialkylaminomethyl aminal derivatives have been
reported by KATRITZKY et al. (69).

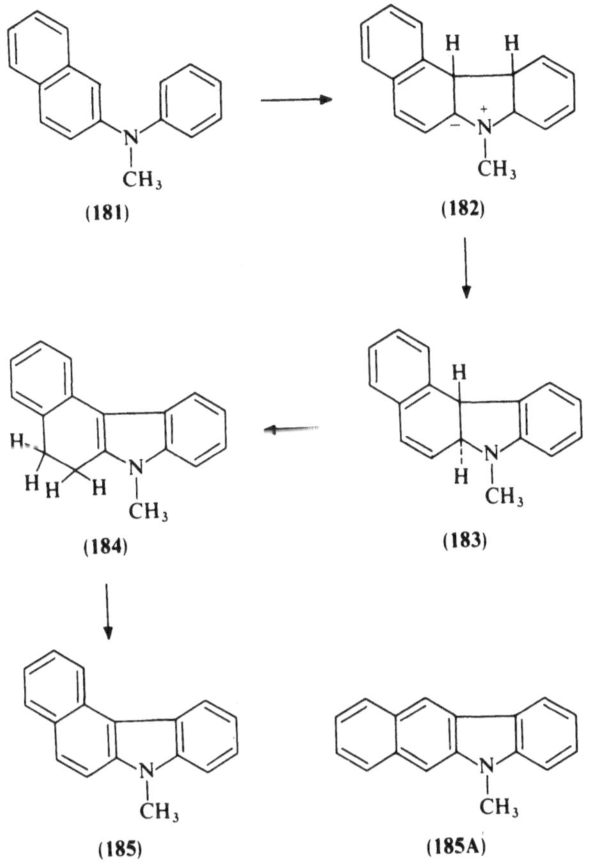

(181)   (182)

(184)   (183)

(185)   (185A)

III. Biogenesis of Carbazole Alkaloids

The isolation of several 3-methylcarbazole derivatives from higher plants (*11*) and the isolation of carbazole (1) from *Glycosmis pentaphylla* (*24*) show that the aromatic methyl group can be eliminated oxidatively from the key compound 3-methylcarbazole (**62**) *via* $-CH_2OH$, $-CHO$, and $-COOH$ functionalities (*18*). The peroxidase induced biological oxidation and elimination of the methyl group attached to the carbazole nitrogen through N-carbinol-amine- and N-formylcarbazole may be cited as a close biogenetic analogy (*71*). Isolation of the dimeric carbazole (**156**) in the course of biomimetic experiments led to the prediction that dimeric alkaloids in higher plants might occur (*19, 106*), a prediction which has been proved to be correct after the isolation of murrafoline (**33**) and other bisalkaloids from *M. euchrestifolia* by Furukawa *et al.* (*81, 44, 45*).

2-Methylcarbazole appears to be the common structural feature in the conventional tricyclic carbazole fragment of many alkaloids isolated from lower plants. Like 3-methylcarbazole in higher plants it could be conceived as a progenitor of the alkaloids occurring in the lower plants (*21*) but Nakamura *et al.* (*124*) have shown in biosynthetic experiments that tryptophan is the progenitor of carbazomycin B (**5**). From the occurrence of ketoanhydro-kinamycin (**186**) in *Streptomyces murayamaensis* formation of the tertiary methyl and hydroxyl on C-1 of carbazomycins G and -H through intermediacy of an epoxide (**187**) might be conjectured.

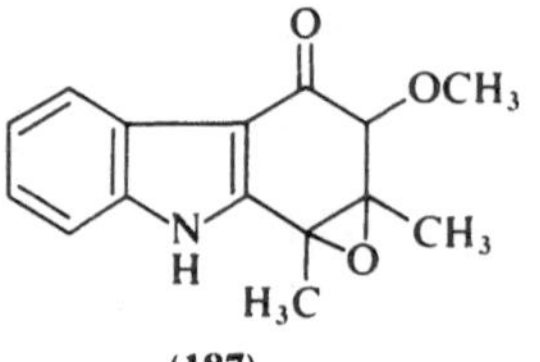

(187) 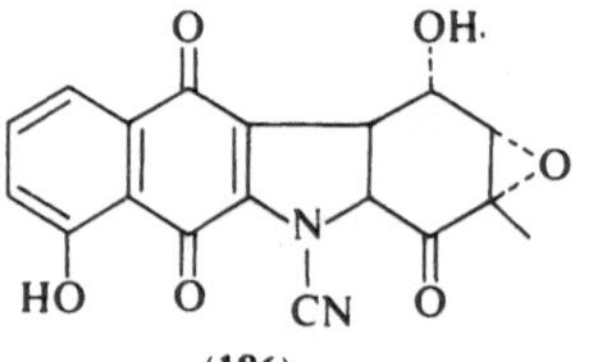   (186)

A decaketide (**188**) cyclising to a benzoanthraquinone (**189**) has been shown to be precursor of the kinamycin group of alkaloids (*111, 112*). The benzoanthraquinone undergoes an unusual process involving oxidation/nitrogen insertion/ring contraction to yield prekinamycin (**190**).

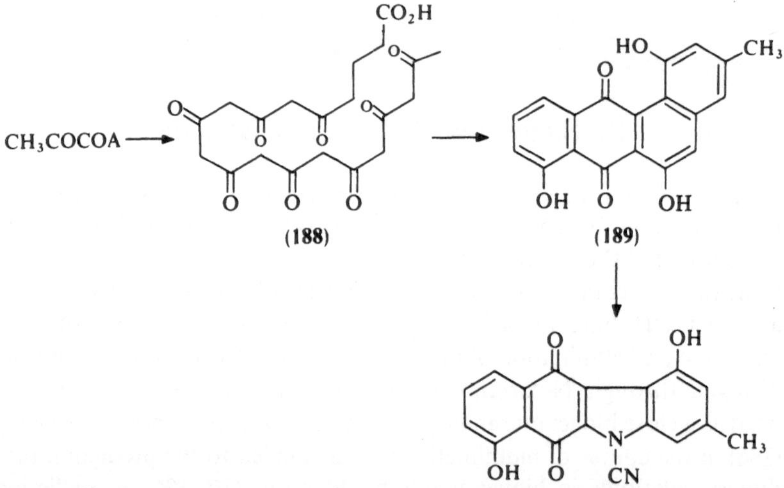

(188)   (189)

(190)

Perhaps more direct proof for this hypothesis will be forthcoming. It is interesting to note that the first natural product with a cyanamide functionality, i.e. N-cyano-*sec.*—pseudostrychnine isolated from *Strychnos wallichina* steud. ex DC, occurs (*14*) in the leaves of the plant together with its N-methyl analogue.

Tubingensins A (**20**) and B (**21**) are considered to arise from indolic diterpenes such as nominine (**191**) which occur in related species of *Aspergillus*.

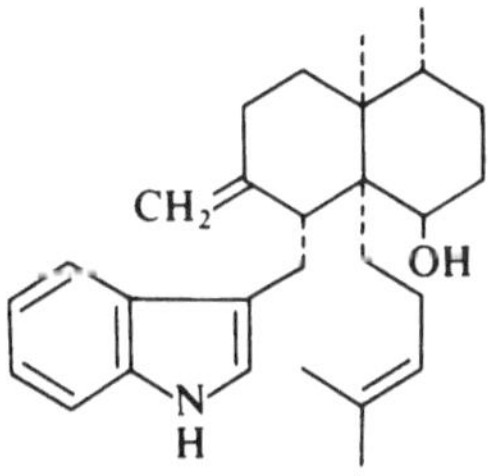

(191)

IV. Biological and Therapeutic Properties of Carbazoles Alkaloids and Congeners

Some of the important biochemical and biological properties have already been mentioned in previous reports (*18, 20, 11*). Additional such properties are reported here.

(a) *Antibiotic activity*: Demethylated glycozoline (**192**) was the most active amongst alkaloids of higher plants tested against *Trychophyton* sp. (10 mg/ml), while carbazomycin B was found to be the most active against *Glomerella cingulata* No 3 and *Elsinoe fawcetti* (3 μg/ml). It was also active against *Trychophyton asteroids* 429 *and T. mentagrophities* 833 (12.5 μg/ml). The phenolic compounds are more active than their O-methyl congeners. Carbazomycinal (**11**) and 6-methoxycarbazomycinal (**24**) inhibit aerial mycelium formation (*74*). The antibiotic activities of kinamycins are much more pronounced (*89*) and increase with decreasing number of acetoxy groups in the order of kinamycins C, A, D, and B (**42, 193–195**). The minimum inhibitory concentrations of kinamycin B are 0.012 μ/ml against *Bacillus subtilis* PC I-219, *B. anthracis*, *Staphylococcus aureus*, *S. albus* and 0.09 to 0.19 against *Vibrio coma*. Derivatives of deacetylkinamycin C show marked activity against *Mycobacterium* ATCC 607 and gram negative organisms like *Escherichia coli* NIHJ, *Klebsiella pneumoniae* and *Shigella sonnei*. On the other hand, derivatives of diacetylkinamycin C had decreased activity against gram positive bacteria and showed no activity against gram-negative bacteria. The sodium periodate oxidation product (196) of deacetylkinamycin C showed markedly strong activity against gram-positive and gram-negative bacteria.

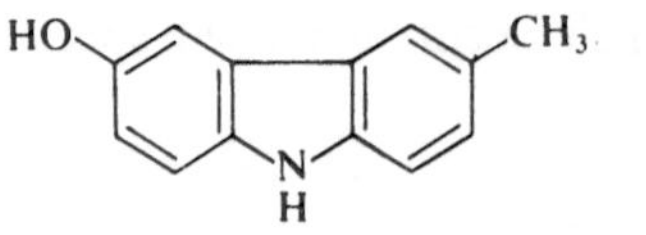

(192)

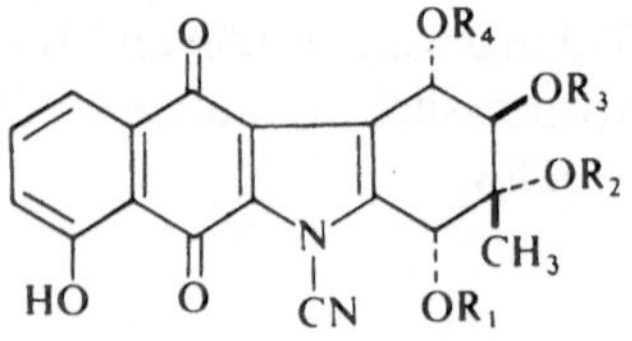

(42) $R_1 = R_3 = R_4 = Ac$; $R_2 = H$
(193) $R_1 = R_2 = R_3 = Ac$; $R_4 = H$
(194) $R_1 = R_3 = Ac$; $R_2 = R_4 = H$
(195) $R_2 = Ac$; $R_1 = R_3 = R_4 = H$

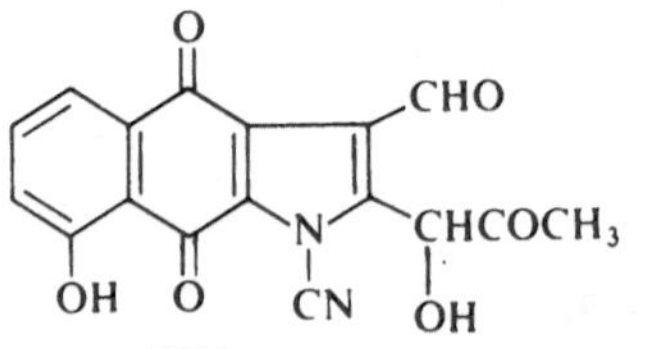

(196)

Some aminoacyl carbazoles have been found to have low antimicrobial properties (36). Addition of 0.01–2.0% by wt. of alkylcarbazolesulphonic acid (197) to bactericides increases the degree to which bacterial growth is prevented (72).

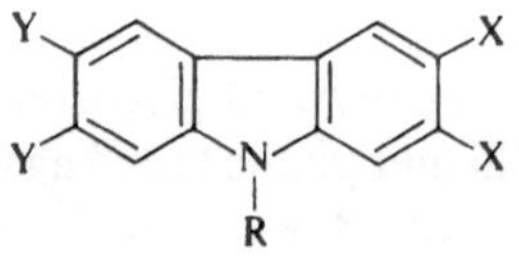

(197) R = Alkyl
X = H/SO$_3$H; Y = SO$_3$H

(b) *Anticancer activity* (104): Weak antitumor activity was exhibited by kinamycin C (42) against Ehrlich ascites carcinoma and against sarcoma-180 (110). Tubingensin A (20) and tubingensin B (21) are cytotoxic to HeLa cells (120), the latter compound being more reactive (IC$_{50}$ 4 µg/ml) than the former one (IC$_{50}$ 23 µg/ml).

(c) *Antiviral activity*: Tubingensin-A (20) and tubingensin-B (21) exhibited activity in an *in vitro* assay against herpes simplex virus type-1 with an IC$_{50}$ of 8 µg/ml and IC$_{50}$ of 9 µg/ml respectively. Carpofen (198) enhanced interferon (IFN) produced by suboptimal concentration of 10-carboxymethyl-9-acridanone (CMA) in murine cell cultures. Carpofen caused an approximately 500 fold increase in CMA induced IFN production in pure bone marrow derived in macrophase culture. Its activity is dependent upon the inhibition of cyclooxygenase (*117*).

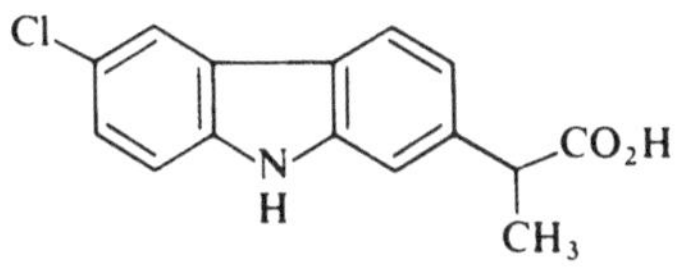

(198)

Several 9-ethyl-substituted bisbasic carbazoles of the general formula (199) were tested against *Encephalo myocarditis*, virus infection in mice (*35*). Compounds with R = Et, R_1 = COOR$_2$ and those with R_2 = ethyl to hexyl have the highest activity. Some basic esters and amides of carbazole have also been tested against the virus. Compound (200) at a concentration of 50 mg/kg produces a longer survival time than the controls (*?*).

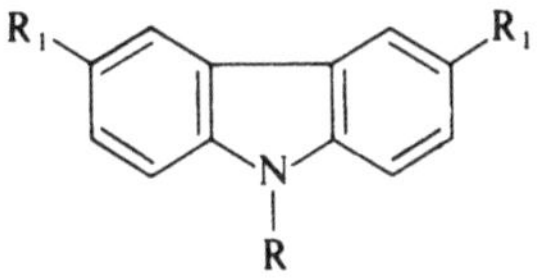

(199)

R = H, Me, Et; R_1 = COOR$_2$/CONMe$_2$/CONEt$_2$/CH
NMe$_2$/CH$_2$NEt$_2$/CHCOOCH$_2$Ph/NHCOOEt; R_2 = Et/Bu/Hexyl

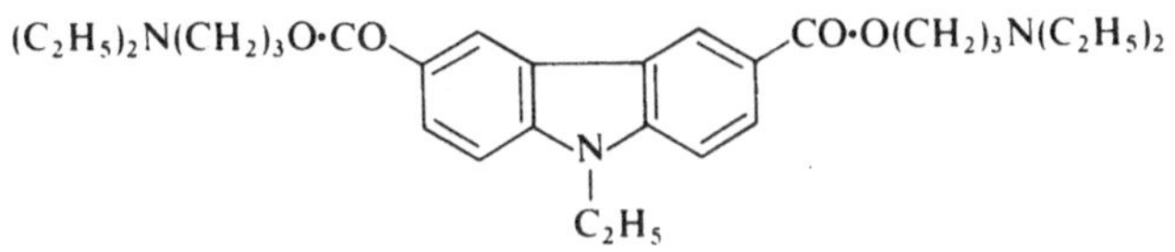

(200)

(d) *CNS and related activities*: Some tetrahydrocarbazoles with alkylamino substitutions in the aromatic ring have been found to produce behavioral changes in animals (*114*).

Compound (201) gives rise to abnormal behavioral pattern while (202) does not and (203) has been claimed to effect temporary improvement in the condition of acutely disturbed patients. Some tetrahydrocarbazoles (204) were shown to inhibit gastric secretion (*14a*).

Pharmaceuticals containing 1,2,3,4-tetrahydro-3-[(1-H-imidazolyl)-methyl]-4-H-carbazole-4-one derivatives have been found to control vomiting and gastric evacuation (*50*). Some ketotetrahydrocarbazoles

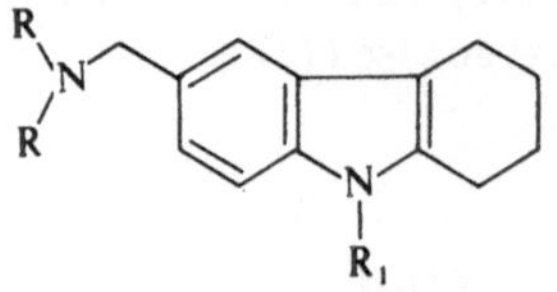

(201) R = CH₃; R₁ = CH₂Ph
(202) R = CH₃; R₁ = H.

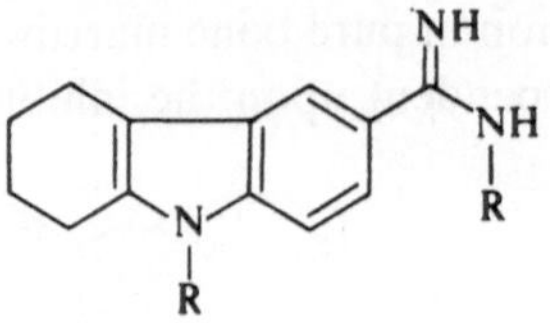

(203) R = Ph; R₁ = H

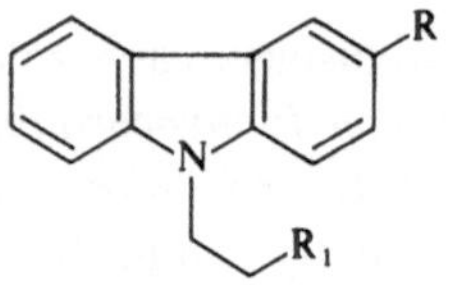

(204) R = Me/Et/CHMe₂/CMe₃/CH₂CMe₃
R₁ = NMe₂/NEt₂/1-pyrrolidinyl/1-piperidyl

have been found to be effective for treatment of emesis, antianxiety and/or hypertensive colic syndrome. Substituted (imidazolylmethyl)-carbazolones having the following general structural pattern (205) have been shown to be 5-HT antagonistic (6, 25, 26).

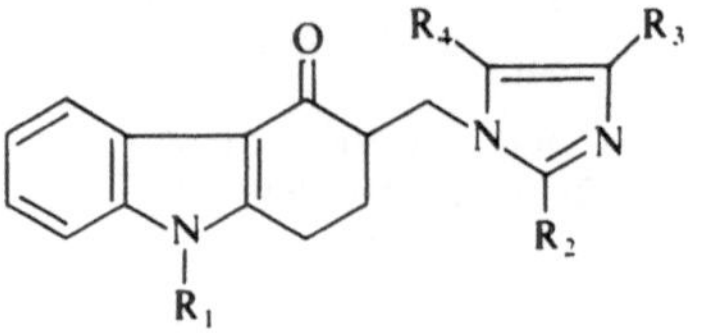

(205)

(a) R₁ = R₂ = R₃ = R₄ = H. (b) R₁ = SO₂Me. R₂ = Me. R₃ = R₄ = H
(c) R₁ = C₃₋₇/Cycloalkyl. C₃₋₁₀ alkynyl: R₂ = H/alkyl/cycloalkyl;
R₁ = R₄ = H, alkyl

Some of them are useful in the treatment of psychotic disorders, anxiety, pain, gastrointestinal dysfunction associated with dyspepsia, peptic ulcer, reflex esophagitis, flatulence (50). Cyclindole (206) and flucindole (207) are neuroleptic agents (77, 123). Cyclindole shows anti-depressant activity while flucindole shows antipsychotic activity.

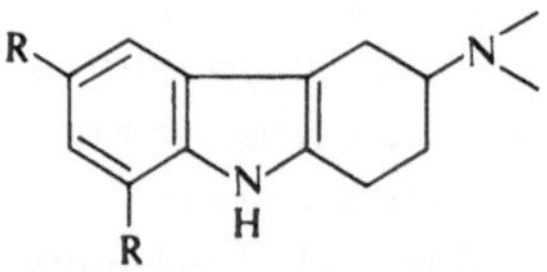

(206) cycliindole. R = H
(207) Flucindole. R = F

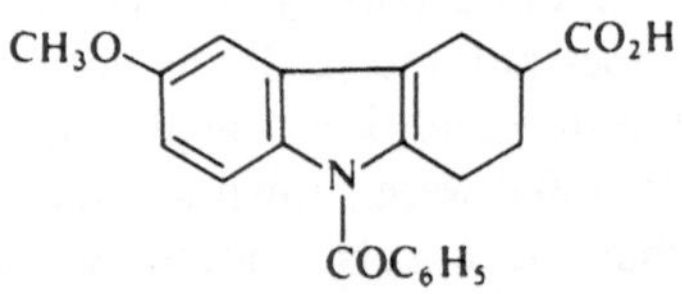

(208)

Oxarbazole (**208**) shows antiallergic activity (*77*). The tetrahydro-carbazole derivatives of the general formula (**209**), R = alkyl, alkoxy, alkyl thio, sulfinyl sulphone, OH, F, Cl, Br, CF_3, cyano, methylene dioxy, Z = CH_2/CH_2–CH_2 are useful central nervous system agents. Above all they posses antiperkinson activity (*55*).

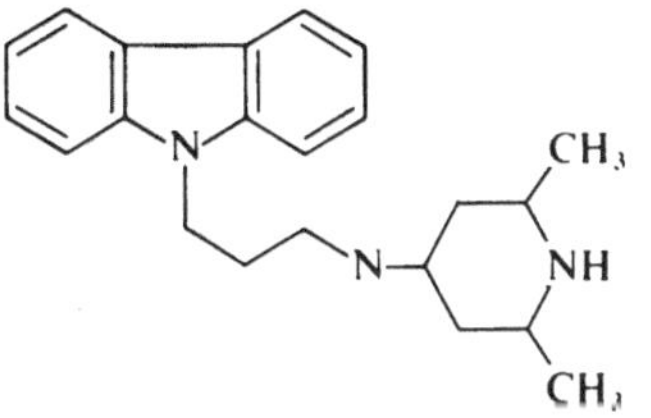

(210)

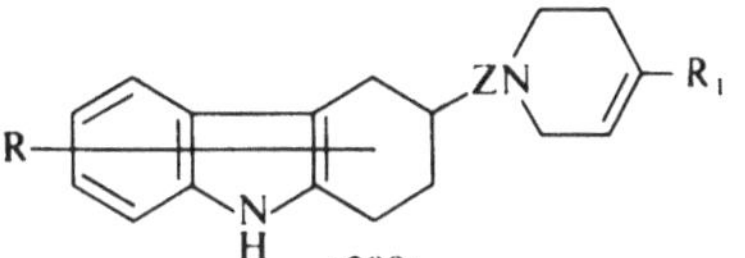

Rincazole (**210**) is a novel antipyretic, neuroleptic agent (*38, 37, 39*).

It has been however found to be specific competitive antagonistic of Sigma sites in brain. It reverses psychotic conditions induced in humans by phencyclidine and/or σ-opiod antagonistics probably by these compound to receptors in brain. Rincazole has indirect effect on dopamine neurons (DA) with relative selection for A10 dopamine cells (*102*).

Murrayazolinine (**211**) has been found to effect sudden lowering of blood pressure in experimental subjects. The data show that the action is not mediated through mascurine H_1 or H_2 receptors (*82*). 3-Chlorocarbazole (**212**) isolated from bovine urine has 'Diazepam like activity' (*76*).

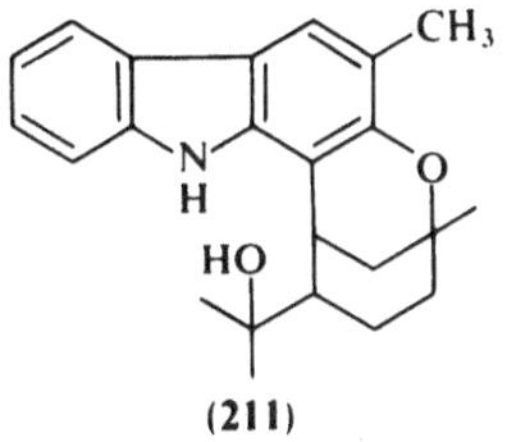

(211) 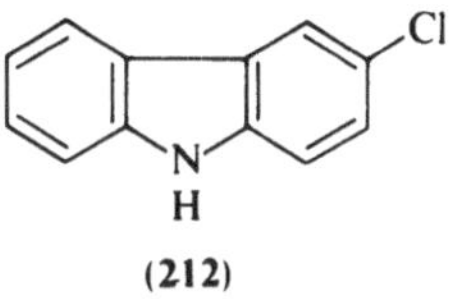   (212)

(e) *Antiinflammatory properties*:   1-Ethyl-8-propyl-1,2,3,4-tetra-hydrocarbazole acetic acid (**213**) is a powerful antiinflammatory compound. Several 2,3,4,9-tetrahydrocarbazole-1-acetic acid (**214**) and their salts are useful as analgesic and antiinflammatory agents.

The carpofen (**198**) group of antiinflammatory substances (*80*) have been shown to have phototoxic properties involving skin photosensitivity (*78*). They induce uv dependent histamine and leucotriene release in peripheral human leucocytes. *In vitro* cellular effect of corpofen was

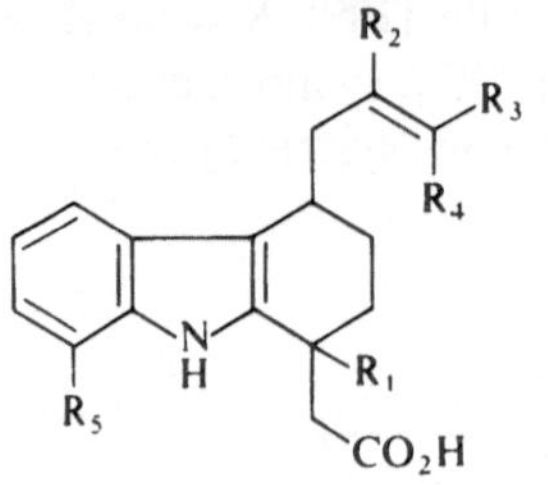

(213) R_1 = alkyl; $R_2 = R_3 = R_4$ = H/alkyl
        $R_2 = R_4 = CH_2 = CHCH = CH_2$;
        R_5 = H/Alkyl/Halo

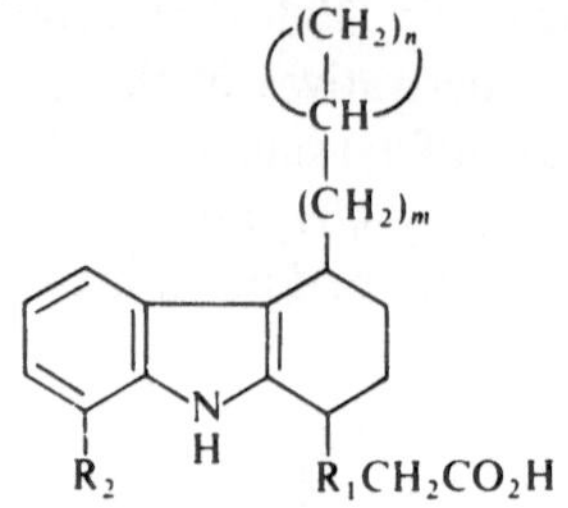

(214) R_1 = alkyl; R_2 = H/
        C_{1-6} alkyl; m = 0, 1; n = 2–5

found to be higher than that of ibuprofen and almost comparable to that of hydrocortisone (*121*). The antiinflammatory activity of carpofen probably involves a concentration dependent inhibition of some neutrophil macrophase function.

(f) *Enzyme inhibitory, anabolic and other effects.* 3-Chlorocarbazole (**212**) is a potent inhibitor of rat liver monamine oxidase (MAO) activity (*33*). Some methyl-substituted tetrahydrocarbazoles have been tested for *in vitro* activity against hydroxylase and dopamine β-hydroxylase inhibitory activity. The potent compounds were evaluated for inhibition of norepinephrine biosynthesis. No significant inhibition was recorded.

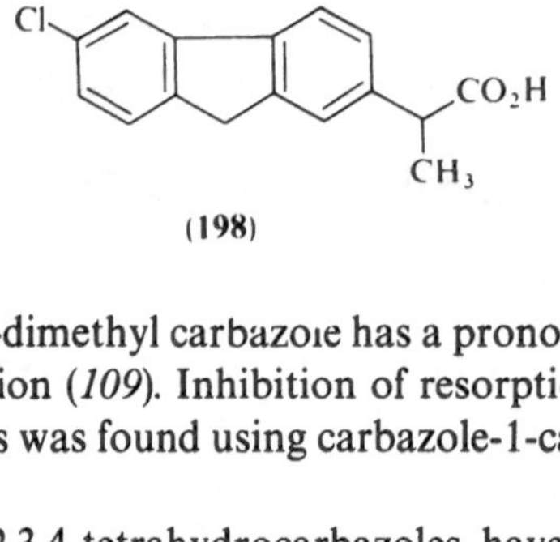

(198)

6-Hydroxy 1,4-dimethyl carbazoıe has a pronounced inhibitory effect on lipid peroxidation (*109*). Inhibition of resorption in organ culture of fetal rat long bones was found using carbazole-1-carboxylic acid as a test material (*103*).

Substituted 1,2,3,4-tetrahydrocarbazoles have been reported to be trypanocidals by Peeca *et al.* (*95*). They have also been found to have low larvaecidal activity (*29*) and some pesticidal properties (*28*).

Tubingensin A (**20**) has moderate and tubingensin B (**21**) has low activity against the widespread crop pest *Heliothis zea*. Carbazole N-carboxamide (**215**) has been found to have growth inhibitory activity (*68*).

1,10-Bis(6-methyl-5H-benzo-[b]-carbazol-11-yl)-decane (**216**) has been found to be a potential bifunctional DNA intercalating agent (*99*).

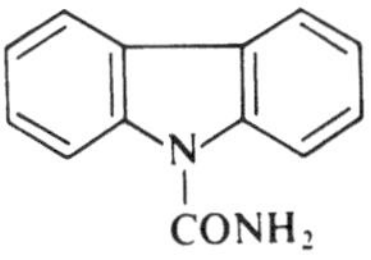

(215) 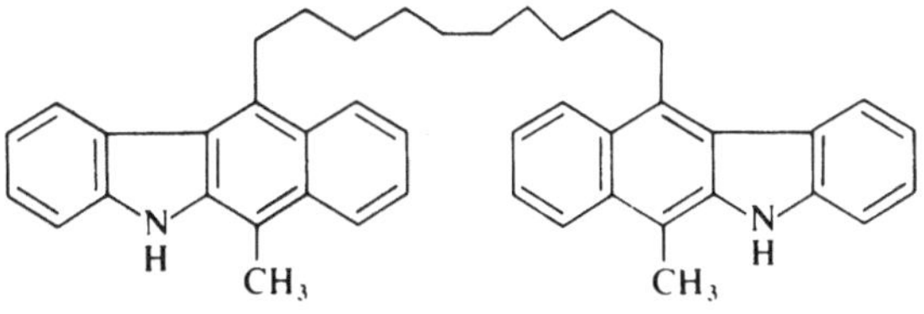   (216)

V. Chemistry of Carbazole Alkaloids

A. Alkaloids from Higher Plants

(i) C_{13}-Alkaloids

1. Murrayastine

Murrayastine (13), $C_{16}H_{17}NO_3$ (M$^+$ 271.1191) was isolated from the dried bark of *Murraya euchrestifolia* as a syrupy liquid by FURUKAWA *et al.* (*41*). Its uv absorption spectrum (λ_{max} 224, 247, 255, 298, 332 and 336 nm) and ir spectrum showed the presence of carbazole skeleton. ^{1}H-NMR data showed the presence of one aryl methyl (δ 2.47), and three aryl methoxy (δ 3.95, 3.96 and 4.00) groups. The *meta*-coupled H-4 (δ 7.32) and H-2 (δ 6.65) signals and their enhancements after irradiation of the aromatic C-methyl showed that C-2 and C-4 were unsubstituted and that the methyl occupied position 3. Presence of the *ortho* coupled signal of H-5 ($J = 8$ Hz) showed that both C-5 and C-6 were unsubstituted. Hence murrayastine could be formulated as 1,7,8-trimethoxy-3-methyl carbazole which has been confirmed by synthesis (*vide* p. 97).

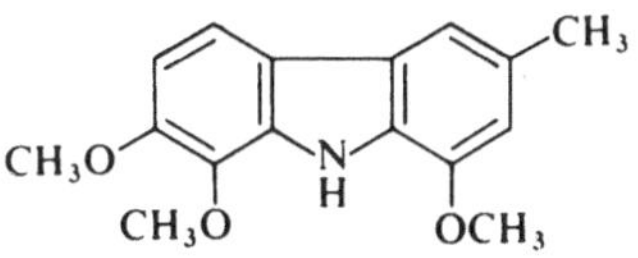

(13) 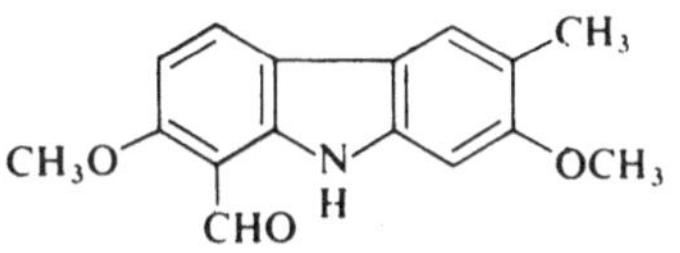   (14)

2. Murrayaline

Murrayaline (14) $C_{16}H_{15}NO_3$ (M$^+$ 269.1002), m.p. 248–50° was isolated as a pale yellow prism from the bark of *M. euchrestifolia* (*41*). Its uv absorption spectrum and ir data showed it to be a carbazole derivative with a hydrogen bonded formyl group. (v_{max} 1650 cm^{-1}) which was also discernible in its ^{1}H-NMR spectrum which showed signal for one aryl methyl (δ 2.35) two methoxy groups (δ 3.90 and 3.98), singlets for H-4 (δ 7.95) and H-1 (δ 6.91) protons and *ortho* coupled H-5 (δ 8.02, *d*, $J = 9$ Hz) and H-6 (δ 6.77 *d*, $J = 9$ Hz) protons. Irradiation of the aromatic methyl signal produced enhancement of the H-4 signal. Irradiation of the methoxy group signal resulted in enhancement of the H-1 and H-6 protons showing that the methoxyls were on C-2 and C-7. These data were consistent with formulation of murrayaline as 3-methyl-2,7-dimethoxy-8-formylcarbazole which has been confirmed by synthesis (*vide* p. 97).

3. O-Demethylmurrayanine

O-Demethylmurrayanine (**217**) $C_{13}H_9NO_2$ (M$^+$ 211), m.p. 237–39° was isolated from stem and root bark of *Clausena anisata* (*85*). The uv spectrum (λ_{max} 226, 244, 255, 278, 291, 336, 346 nm with log ε 4.40, 4.51, 4.39, 4.59, 4.45, 4.22, 4.22) showed it to be a 3-formyl carbazole derivative. This and colour reactions showed it to be a phenolic carbazole. The ^{1}H-NMR spectrum indicated the presence of deshielded metacoupled H-4 and H-2 protons, the four protons of the ring A and the proton of the CHO group (δ 9.89). Hence the alkaloid was formulated as 1-hydroxy-3-formyl-carbazole. The ^{13}C-NMR spectrum also support the structure.

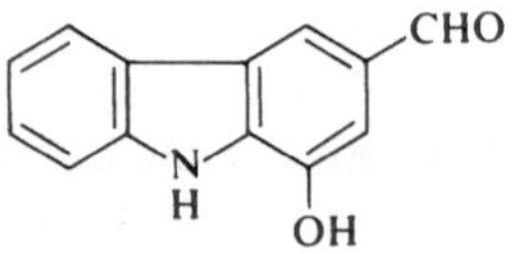

(217)

4. 3-Formylcarbazole

3-Formylcarbazole (**218**) $C_{13}H_9NO_2$ (M$^+$ 195.0683) was obtained as a colourless oil from *Murraya euchrestifolia* (*59*). The uv spectrum was characteristic for a 3-formylcarbazole which was also supported by the ir and ^{1}H-NMR spectra.

5. N-Methoxy-3-formylcarbazole

N-Methoxy-3-formylcarbazole (**219**) $C_{14}H_{11}NO_2$ (M$^+$ 225.079) (oil) obtained from *M. euchrestifolia* (*59*) had a uv absorption spectrum (λ_{max}

236, 272, 288, 320 nm) very similar to that of 3-formylcarbazole which was also supported by the ir spectrum (1680 cm^{-1}). The ^{1}H-NMR spectrum showed the presence of all protons of the carbazole skeleton except that of H-9, which was replaced by a signal for a methoxy group (δ 4.27). These data led to structure (219) which was also supported by its ^{13}C-NMR spectrum. The mass spectral data showed peaks at m/z 194 (M-31), 166 (M-31-28) for the loss of methoxy and formyl groups respectively.

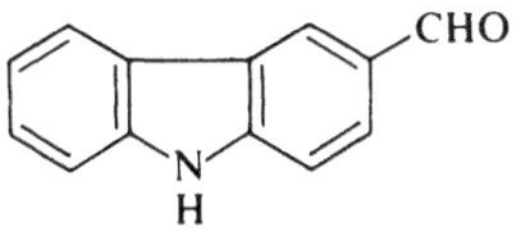

(218) 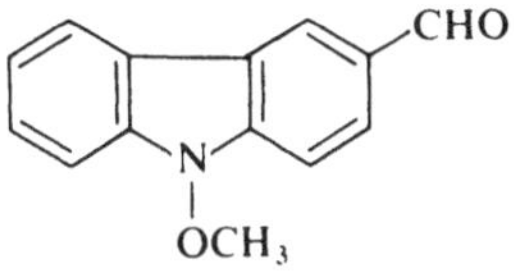   (219)

6. Carbazole

Carbazole, $C_{12}H_9N$, m.p. 225° (1) was isolated from *Glycosmis pentaphylla* Retz DC (24). It was characterised by its ir, uv and mass spectral data and direct comparison with an authentic specimen of carbazole.

(ii) C_{18}-Alkaloids

1. Dihydroxygirinimbine

Dihydroxygirinimbine (220), $C_{18}H_{19}NO_3$ (M$^+$ 297.1365), m.p. 189–90°, $[\alpha]_D^{\text{methanol}} - 4.0$ was isolated from the root bark of *Murraya euchrestifolia* (43). The ir, and the uv absorption spectrum (λ_{max} 215, 238, 254, 259, 303, 332 nm) showed it to contain the 3-methylcarbazole skeleton with hydroxyl functions. Presence in the ^{1}H-NMR spectrum of an isolated 4-spin system (δ 7.06, 7.24, 7.50, 7.96) showed the absence of substitution in ring A while the presence of an H-4 singlet (δ 7.72) and the aryl methyl at C-3 (δ 2.28) showed that position 2 was substituted. ^{1}H-NMR data further showed oxygen-linked carbon carrying a *gem*-dimethyl group (δ 1.24 and 1.52) and a pair of doublets at δ 3.79 and 4.92 (J = 8 Hz each) which shifted to δ 5.8 and 6.22 on conversion to the diacetate, m.p. 159–161° (221). The data were consistent with the presence of a 2′,2′-dimethyldihydroxypyranocarbazole system, the hydroxyls being *trans* to each other. Hence the alkaloid could be formulated as dihydroxygirinimbine (**220**). The isolation of a compound identical with (**220**) in all respects from m-chloroperbenzoic acid oxidation product of girinimbine (**3**) along with isomeric *cis*-diol (**222**) confirmed the *trans* diol structure of dihydroxygirinimbine. The *cis* diol (**222**) was also obtained as a major product on OsO_4 oxidation of (**3**).

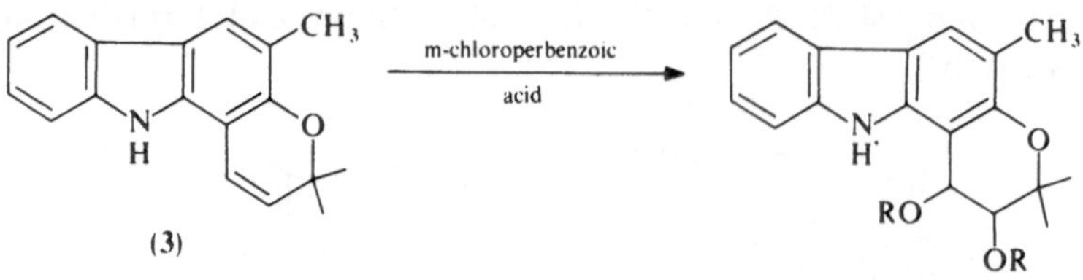

(**3**)

m-chloroperbenzoic
acid

(**220**) R = H (trans)
(**221**) R = COCH$_3$
(**222**) R = H (cis)

2. Pyrayafoline

Pyrayafoline (**121**), $C_{19}H_{19}NO_2$ (M$^+$ 293.1399), m.p. 228–31° was obtained as colourless plates from the root bark at *M. euchrestifolia* (*41*). Its uv spectrum (λ_{max} 222, 239, 286(sh), 295, 334 nm) and ir data showed it to have a DMP system fused to a carbazole skeleton which was also substantiated by a mass spectral peak at m/z 278 (M-15). The ^{1}H NMR spectrum had signals for aromatic methyl (δ 2.33), aromatic methoxy (δ 3.89) and characteristic signals for a DMP system (δ 1.47 for 6H and one proton doublet at δ 5.67 and 6.58 (J = 10 Hz each). In addition, it had signals for H-4 (δ 7.62, s), H-5 (δ 7.63, J = 8 Hz) and H-6 (δ 7.77, br s). These data were consistent with formula (**121**) for pyrayafoline which has been confirmed by synthesis through the diphenylamine route (*vide* p. 105).

3. Isomurrayafoline B

Isomurrayafoline B (**223**), $C_{19}H_{21}NO_2$ (M$^+$ 295.1567), m.p. 158–61° was isolated from *M. euchrestifolia* (*58*). The uv spectrum (λ_{max} 213, 237, 264, 310, 330 sh nm) showed the presence of a carbazole skeleton. The ^{1}H NMR spectrum showed the presence of a DMA chain [δ 1.74, (3H, s), δ 1.88, (3H, s), δ 5.30, (1H, t, J = 7 Hz), δ 3.60 (2H, d, J = 7 Hz)], a methoxy group (δ 3.90) and an aryl methyl group (δ 2.38); one OH (δ 4.74, 1H, br, s) and NH proton (1H, br, s). Signals for H-5 and H-4 appeared at δ 7.70 (J = 8 Hz) and 7.67 (s) respectively. Enhancement of the H-4 signal

by irradiation of the aromatic methyl and that of H-6 by irradiation of
the methoxy group suggested that C-2, C-3, C-7, and C-8 were substituted. Hence structure (223) was suggested for isomurrayafoline B.

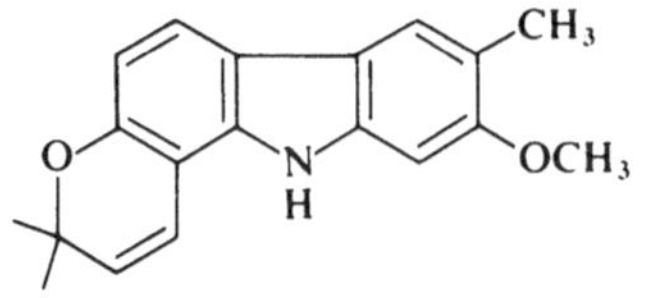

(121) 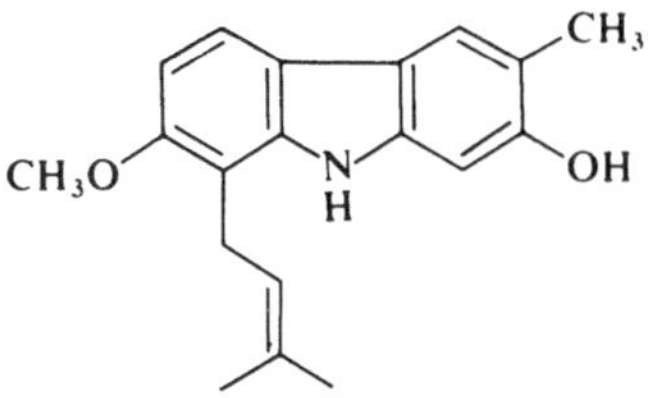   (223)

4. Atanisatin

Atanisatin was isolated from *Clausena anisata* along with clausanitin
(56) and has been assigned structure (224).

The C_{18}-alkaloid, (225) and (226) have recently been reported from
M. euchrestifolia (60).

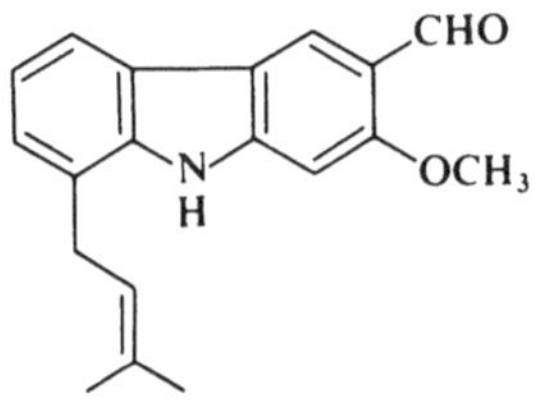

(224) 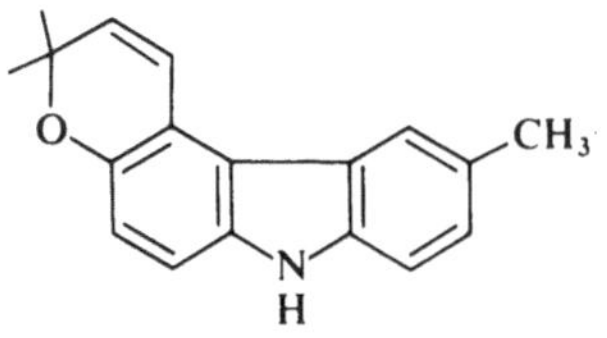   (225)

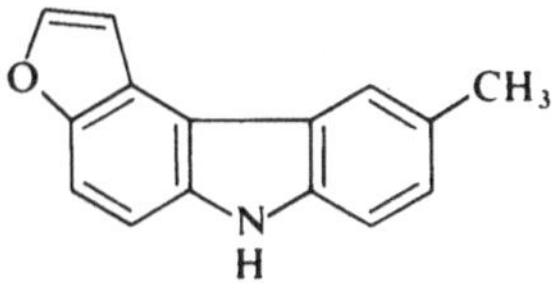

(226)

(iii) C_{23}-Alkaloids

1. Murrayazolinol

Murrayazolinol (227), $C_{23}H_{22}NO_2$ (M$^+$ 347), m.p. 290°, $[\alpha]_D^{CHCl_3}$
$\pm$ O was isolated from *Murraya koenigii* Spreng (10). Its uv spectrum

(λ_{max} 245, 265, 305 nm with log ε 4.85, 4.32, 4.40) showed the presence of a 2-oxygenated carbazole chromophore and the ir spectrum showed the presence of hydroxyl or NH function (ν_{max} 3360). The ^{1}H-NMR spectrum showed the presence of a *gem*-dimethyl group (δ 1.93, 1.5, 3H each s), a benzylic proton (δ 3.33, 1*H*, br) and the carbinyl hydrogen (δ 3.8 dd $J = 7.2$ Hz). Also present were signals for H-4 (δ 7.50, s), H-5 (δ 7.95), H-6 and H-7, (δ 7.15, 7.40 m). The ^{1}H NMR spectrum was very similar to that of murrayazoline lacking the signal for the NH proton. The mass spectrum showed the molecular ion at m/z 347; other ionic species were at m/z 332, 314, 248, 120. In consideration of the mass spectral data, a comparison of the benzylic proton signals (δ 3.33) of murrayazolinol with those of murrayazolinine (δ 3.68) and isomurrayazoline (δ 3.08) (9) and the spectral pattern of the carbinyl hydrogen signal at δ 3.80, the hydroxyl group was placed on a carbon adjacent to the oxygen linked carbon atom of the pyran ring; thus structure (**227**) was advanced for murrayazolinol.

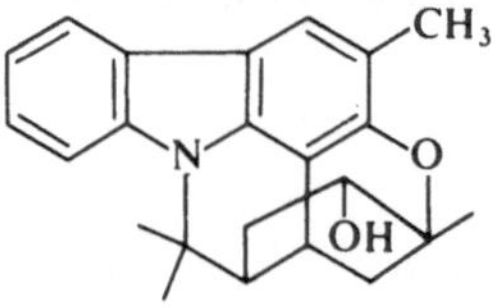

(**227**)

(iv) Carbazoloquinones

Some carbazole alkaloids which share a 1:4 quinone system with a methyl group at C-3 (*46*) have been discussed in this section irrespective of whether they possess a C_{18} or C_{23} skeletal unit. In the previous review structures of murrayaquinones A, B, C and D have been discussed (*11*).

1. Pyrayaquinone A

Pyrayaquinone A (**25**), $C_{18}H_{15}NO_3$ (M$^+$ 293.1053), m.p. 222° was isolated from the stembark of *M. euchrestifolia* (*42*). The uv (λ_{max} 220 sh, 252, 308 sh, 460 nm) and ir (ν_{max} 1660, 1640, 1610 cm^{-1}) spectra show that it has a carbazoloquinone system. The presence of a DMP system fused to a carbazoloquinone fragment was discernible from its ^{1}H-NMR signals (δ 1.48, 6H, s; doublets for one proton each at δ 5.72 and 6.48 with $J = 10$ Hz). The doublet at δ 6.46 and a 3H signal at δ 2.16 for a methyl at C-3 show no substitution at C-2. The singlets H-5 and H-8 protons show that the DMP system is fused to ring A in a linear fashion with oxygen at C-7. The high intensity mass spectral peak at m/z 278 also supports this tetracyclic system in pyrayaquinone-A. The structure has been confirmed by synthesis through the diphenylamine route (*vide* p. 105).

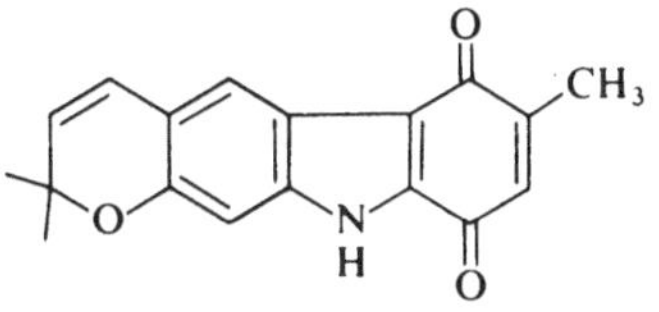

(25)

2. Pyrayaquinone B

Pyrayaquinone B (**26**), $C_{18}H_5NO_3$ (M^+ 293.1052), m.p. 244°, was isolated from *M. euchrestifolia* (*42*). Like pyrayaquinone A its uv (λ_{max} 229 sh, 248, 295 sh, 320, 410 nm) and ir (ν_{max} 1650, 1640, 1635, 1605 cm^{-1}) spectra showed it to contain a 1,4 carbazoloquinone system. The presence of the DMP system fused to a carbazoloquinone has been deduced from the ^{1}H NMR spectrum (δ 1.48, 6H, one proton doublets at δ 5.70 and 6.60 with $J = 10$ Hz each). The quartet for 1 H at δ 6.44 and the singlet at δ 2.14 with long range coupling could be reconciled with the presence of a methyl at C-3 and no substitution at C-2. The *ortho*-coupled signals for H-5 and H-6 (δ 7.94 and 6.87, $J = 9$ Hz) show that the DMP system is fused to the carbazoloquinone system, in an angular fashion with the oxygen at C-7. The high intensity mass spectral peak at m/z 278 also supports the presence of the tetracyclic carbazoloquinone system. The structure has been confirmed by synthesis through the diphenylamine route (*vide* p. 105).

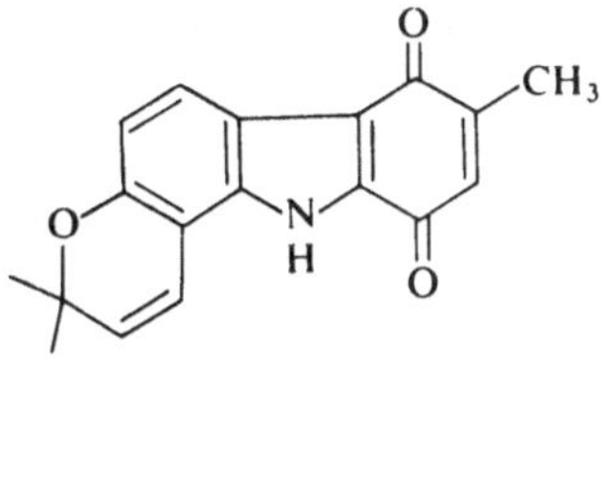

(26) 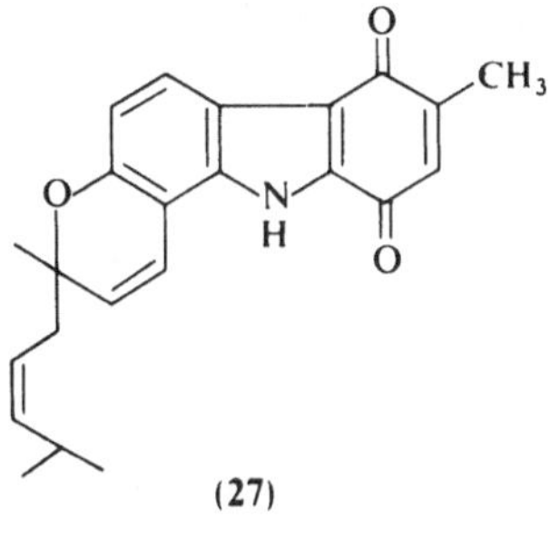   (27)

3. Pyrayaquinone C

Pyrayaquinone C (**27**), $C_{23}H_{23}NO_3$ (M^+ 361.1666), m.p. 223° was isolated from *M. euchrestifolia* (*59*). Its uv (λ_{max} 248, 275 (sh), 294, 392 nm) and ir (ν_{max} 1640, 1600 cm^{-1}) spectral data were similar to those of other 1:4 carbazoloquinone alkaloids. The ^{1}H-NMR signals for the aryl methyl group (δ 2.15, d, $J = 1.7$ Hz) and the olefinic proton doublet at δ 6.45 (1H $J = 1.7$ Hz) suggested that like other carbazoloquinones it had a methyl at C-3. Doublets (δ 5.68 and δ 6.63 with $J = 10$ Hz each) and the singlet for a methyl group (δ 1.44) showed the presence of a dimethyl-pyranocarbazole in which one methyl has been replaced by a chain of six carbons similar to mahanimbine. The presence of *ortho*-coupled H-5 (δ 7.94) and H-6 (δ 6.86) showed that the pyran ring was fused to the carbazoloquinone fragment at the 7, 8 position. This was also supported by the presence of a quinonoid carbazolopyrilium ion at m/z 278 as in other pyranocarbazoloquinone alkaloids. From these data structure (**27**) was proposed for pyrayaquinone C.

(V) Dimeric Carbazole Alkaloids

The first dimeric carbazole alkaloid, murrafoline, and all other dimeric alkaloids have so far been isolated from *Murraya euchrestifolia* Hayata by FURUKAWA *et al.* (*81*). These alkaloids are built up from monomeric alkaloids with previously known carbon skeletons and have $C_{26}(C_{13} + C_{13})$, $C_{31}(C_{13} + C_{18})$, $C_{41}(C_{18} + C_{23})$ carbon skeletons. In the previous review (*11*) murrafoline, bismurrayafoline A and bismurrayafoline B were discussed; other members are discussed here.

(a) C_{26}-Alkaloids

1. Bismurrayafolinol

Bismurrayafolinol (**37**), $C_{28}H_{26}N_2O_3$ (colourless oil) (M^+ 436.1784) was isolated from the root bark of *M. euchrestifolia* (*58*). Its uv (λ_{max} 225, 244, 252, 281, 292, 329, 340 nm) and ir spectrum showed it to be a carbazole alkaloid with a hydroxyl function. From the base peak at m/e 210 ($M^{+ +}$) it was considered to be a dimeric alkaloid. The ^{1}H NMR spectrum showed the presence of two aromatic methoxyl groups (δ 3.83, 3.97), H-2, H-2' (δ 6.78, 7.06), *ortho*-coupled H-5, H-5' (δ 7.92 and 8.07, d, $J = 8$ Hz each) and a singlet for H-4 (δ 7.74). In addition signals from δ 7.10 to 7.50 for seven protons were present. Also signals for two benzylic methylenes, one attached to nitrogen and the other to oxygen (δ 6.01 and 4.86) were discernible. By comparing the ^{1}H-NMR data with those of bismurrayafoline A the structure (**37**) was advanced for bismurrayafolinol which has been confirmed by the synthesis of its acetate (**155**) (*vide* p. 110).

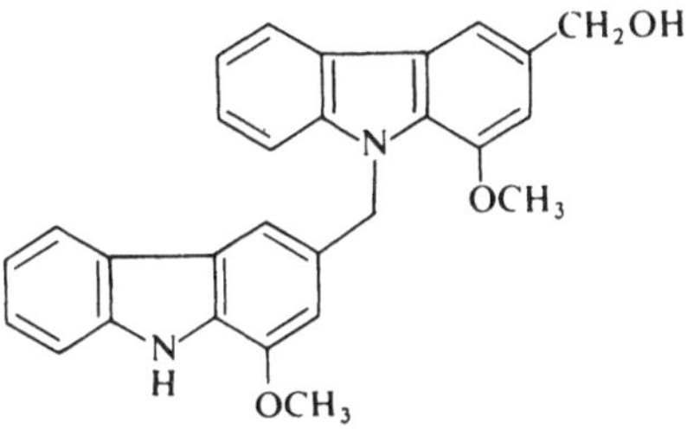

(37) 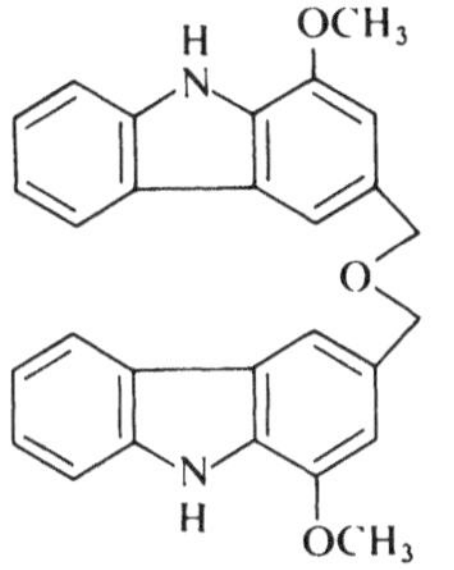   (38)

2. Oxydimurrayafoline

Oxydimurrayafoline (**38**), $C_{28}H_{24}N_2O_3$ (colourless oil) (M$^+$ 436.1809) was isolated from *M. euchrestifolia (58)*. The uv (λ_{max} 226, 242, 253, 260, 280, 291, 324, 337 nm) and ^{1}H NMR data showed the presence of two NH protons (δ 8.27) and two aromatic methoxy groups (δ 4.01). A signal at δ 4.76 (4H, s) was attributed to two benzylic oxymethylene groups. Comparison of ^{1}H NMR data of oxydimurrayafoline with those of murrayafoline A and consideration of the mass spectral peak at m/z 211, about at half the value of the molecular ion peak at m/z 436, indicated the presence of a dimeric alkaloid built up from monomeric units like those of murrayafoline A. This was also substantiated by enhancement of the H-2 signal on irradiation of the aromatic methoxy signal at δ 4.01 while after irradiation of the benzylic methylene enhancement of the H-4 and H-2 signals was observed. From all these data the structure of oxydimurrayafoline was advanced as (**38**).

3. Murrafoline F

Murrafoline F (**228**), $C_{28}H_{24}N_2O_2$ (colourless oil) (M$^+$ 420.181) was isolated from *M. euchrestifolia (59)*. It had a uv absorption spectrum characteristic of a carbazole nucleus. In the ^{1}H-NMR spectrum signals of two methoxy groups (δ 4.14 and 3.93) and an aromatic methyl group (δ 2.38) were readily discernible. The H-H-COSY NMR spectrum showed that a one-proton signal at δ 7.74 was long range coupled to the aryl methyl. It also exhibited two four-spin proton systems and one three-spin proton system indicating the presence of a 2-unsubstituted ring A and a 3-substituted ring C. Methylene proton signals at δ 4.43 (2H, s) and a carbon signal at δ 32.2 in the ^{1}H- and ^{13}C-NMR spectra respectively were due to the methylene bridge. This was supported by the mass spectral fragment at m/z 223. Enhancement of the H-4 signal on irradiation of the methyl signal and enhancements of the H-4′ and H-2′ signals on irradiation of the benzylic methylene showed that H-4, H-4′ and H-2′ were unsubstituted. Irradiation of the methoxy signal did not cause enhancement of any signal showing that one of the methoxy groups was N-bonded and the other at C-1. The intense ionic peak at m/z 390 with loss of a methoxy group and appearance of 12 Sp^2 carbons as doublets in the ^{13}C-NMR spectrum further substantiated that one methoxy group was at nitrogen. From these data structure (**228**) was proposed for murrafolin F.

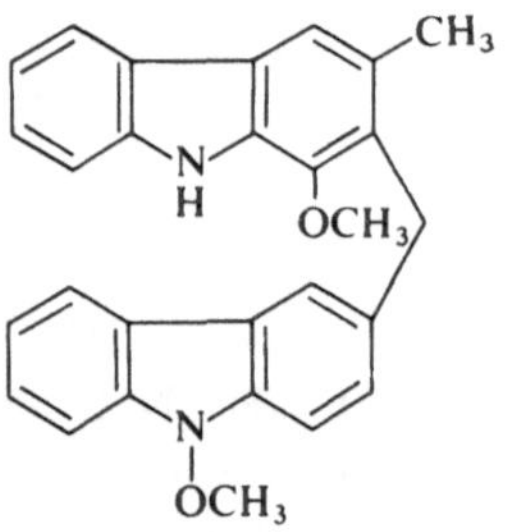

(**228**)

(b) C_{31}-Alkaloids

1. Murrafoline B

Murrafoline B (**19**), $C_{32}H_{30}N_2O_4$ (M$^+$ 474.323) m.p. 234–37° was isolated from *M. euchrestifolia* (*44*). Its uv absorption spectrum (λ_{max} 208, 226, 240, 292, 304, 330 nm) showed it to be a carbazole alkaloid. Its molecular diion peak (M^{++} 237) showed it to be a dimeric alkaloid. The ^1H-NMR spectrum showed it to contain one aryl methoxy (δ 3.87) and two aryl methyl (δ 2.48 and 2.49) groups. The signals at δ 2.30, 2.38 and 4.69 together with the signal for two oxygen linked tertiary methyl groups (δ 1.46 and 1.56) suggested the presence of a 4′-substituted 2′,2′-dimethyldihydropyran ring system. Decoupling experiments showed that it had an unsubstituted and one 8-substituted ring A. Irradiation of the aromatic methyls showed enhancement of signals of H-2′ (δ 6.62) H-4 and H-4′ (δ 7.44 and δ 7.79), irradiation of the methoxy protons on the other hand (δ 3.87) produced enhancement of the H-2′ signal (δ 6.62). These results coupled with the mass spectral fragments at m/z 263 and 211 attributable to dihydrogirinimbine and murrayafoline A respectively were consistent with the assignment of structure (**19**) to murrafoline B. This has been confirmed by synthesis involving the treatment of a mixture of murrayafoline A and girinimbine with Nafion 117 in refluxing aqueous methanol for 48 hrs (*vide* pp. 81, 111).

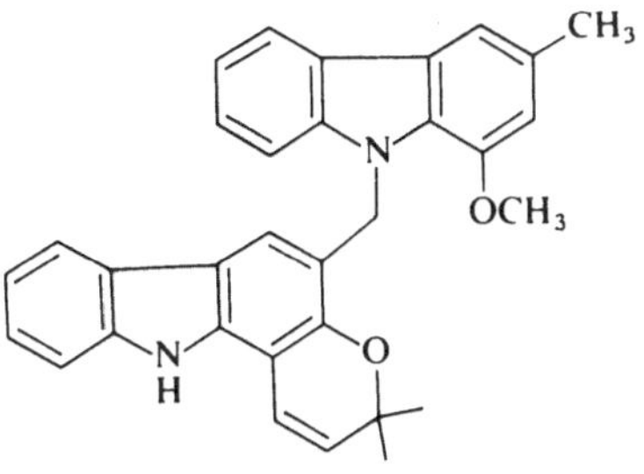

(229) 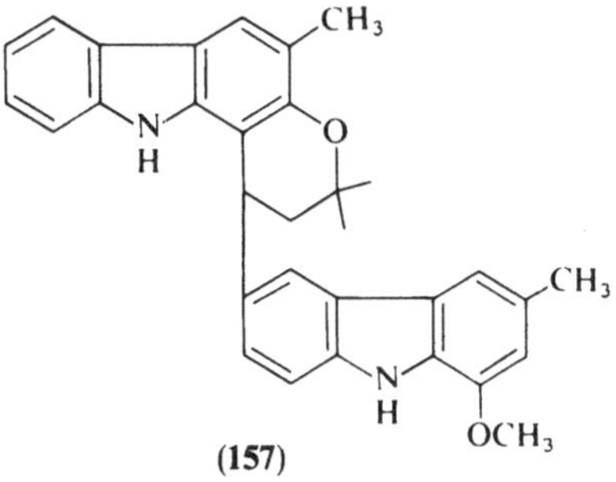   (157)

2. Murrafoline D

Murrafoline D (**157**) isolated from *M. euchrestifolia* (*44*) has been shown to have a structure identical with the synthetic product (**157**) isolated during Nafion 117 catalysed synthesis of biscarbazole (**157** – *vide* p. 111).

3. Murrafoline E

Murrafoline E (**229**) (colourless oil) $C_{32}H_{28}N_2O_2$ (M^+ 420.1831) was isolated from *M. euchrestifolia* (*59*). Uv (λ_{max} 228 (sh), 238, 255 sh, 263 sh, 287, 328, 340, 352 nm) and ir spectra showed it to be a carbazole derivative. The base peaks at m/z 262 and the peak at 211 suggested presence of a binary alkaloid. Its ^{1}H NMR spectrum showed the presence of an aromatic methyl (δ 2.51), a methoxy (δ 3.94), a DMP system [δ 1.44, 6H, s; δ 5.79, 6.92 (1 H, d each, $J = 10$ Hz)] and a benzylic methylene attached to the nitrogen (δ 6.01, 2H, s). The H-H COSY spectrum of murrayafoline E showed the presence of two four-spin proton systems in the aromatic proton region which indicated the presence of two unsubstituted rings A in both fragments. H-H COSY further showed a correlation between a broad one proton singlet at δ 7.58 (H-4) and another at δ 6.90 (H-2) and between a broad one proton singlet at δ 7.34 (H-4') and a two proton singlet at δ 6.01 (benzylic methylene). Enhancement of the H-2 signal due to irradiation of the methoxy signal and enhancements of the H-4 and H-2 signals on irradiation of the methyl signals were observed. Based on these data structure (**229**) was advanced for murrafoline E.

(c) C_{36}-Alkaloids

1. Murrafoline C

Murrafoline C (**34**), $C_{36}H_{34}N_2O_2$ (colourless oil) (M^+ 526.2611) was isolated from *M. euchrestifolia* (*44*). Its ir and uv spectra (λ_{max} 224, 242, 251, 291, 328, 342 nm) showed it to have a carbazole skeleton. Its ^{1}H NMR spectrum showed ABX type (δ 2.27, 2.31 and 4.63) and AB type systems (δ 5.56 and 6.07) as well as the presence of four oxygen linked tertiary methyl groups (δ 1.38, 1.41, 1.42, 1.54) characteristic of 2',2'-dimethyl-Δ³'-pyran and 2',2'-dimethyldihydropyran system. Comparison of the ¹H-NMR data of murrafoline C and murrafoline B suggested the presence of girinimbine and dihydrogirinimbine units linked through C-9' and C-8. The proposed structure was also supported by its spectral fragmentation.

B. Alkaloids from Lower Plants

(i) Carbazomycins

Eight alkaloids of the carbazomycin group have been reported. The structures of these alkaloids were deduced on the basis of the detailed structural studies of carbazomycin B discussed previously (11).

1. Carbazomycin C

Carbazomycin C (39), $C_{16}H_{17}NO_3$ (M⁺ 271.1191), m.p. 198–198.5° was isolated from *Streptoverticillium ehimense* (86). The uv spectrum [λ_{max} 227, 245, 260 (sh), 287 (sh), 295, 341, 354 nm with ε 24900, 24000, 12500, 7600, 12200, 3600, 4200] and the ir spectrum showed it to be a carbazole derivative. The ¹H-NMR spectrum showed the presence of two Ar-methyl groups (δ 2.33 and 2.36), two Ar-methoxy groups (δ 3.74 and 3.84) and one hydroxyl (δ 8.06) as well as signals for H-5 (δ 7.77), H-7 and H-8 (δ 6.91 and 7.31, dd). The ¹³C NMR spectrum also supported these assignments. Due to shifts induced by presence of the methoxy groups, the signals for C-6, C-5 and C-7. C-8 appear at δ 154(s), δ 106.2(d) and δ 113.9(d), δ 111.5 while C-8a appeared at δ 135.6. From these data 1,2-dimethyl-3,6-dimethoxy-4-hydroxycarbazole has been proposed for carbazomycin C.

2. Carbazomycin D

Carbazomycin D (7), $C_{17}H_{19}NO_3$ (M⁺ 285.1374), m.p. 129·5–130° was isolated from the above source. The uv [λ_{max} 229, 247, 255 (sh), 291 (sh), 300, 340, 356 nm with ε 30500, 27800, 19300, 10300, 17000, 4300, 4500] and ir data were similar to those of carbazomycin C. The ¹H-

NMR spectrum was similar to that of carbazomycin C except for the presence of additional methoxy signal which replaced the signal of the hydroxyl group in carbazomycin C. Carbazomycin C on methylation with dimethyl sulphate and alkali gave carbazomycin D. Evidently, carbazomycin D is 1:2-dimethyl-3-4,6-trimethoxycarbazole which is also supported by the ^{13}C-NMR spectrum (86).

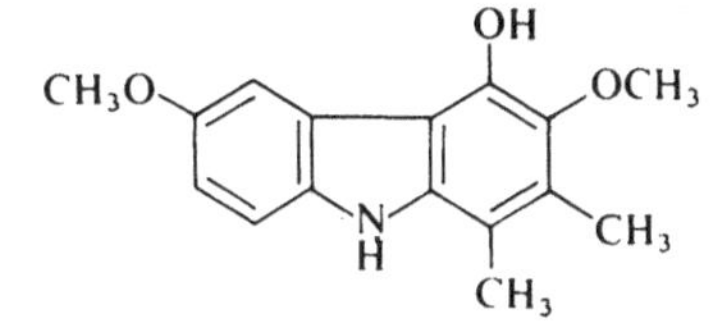

(39) 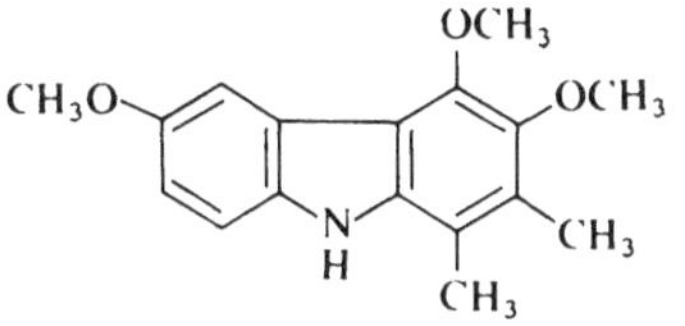   (7)

3. Carbazomycinal

Carbazomycinal (11), $C_{15}H_{15}NO_3$ (M$^+$ 255.0875), m.p. 224° was isolated from a member of the *Streptoverticillium* genus by MARUMO *et al.* (74). From its uv (λ_{max} 214, 227, 263, 295, 320, 372 nm with ε 24 400, 23 200, 1 2200, 16 300, 5100, 8400) and ir (ν_{max} 1660 cm^{-1}) spectra the presence of a 1-formylcarbazole chromophore was readily deduced. The ^{13}C-NMR spectrum contained the signal of a carbonyl group at δ 189.4. In the ^{1}H-NMR spectrum signals due to four neighbouring protons of ring A showed that ring A was unsubstituted. Enhancement of the resonance of aldehydic proton on irradiation of the methyl group showed that the methyl was at C-2 and the aldehyde at C-1. The anisotropic deshielding effect on the signal of additional methoxy group of O-methyl carbazomycinal (δ 3.8) due to ring A showed that the hydroxy group in (11) was at C-4. All this information led to formulation of carbazomycinal as 1-formyl-2-methyl-3-methoxy-4-hydroxycarbazole which is also supported by the ^{13}C-NMR data.

4. 6-Methoxycarbazomycinal

6-Methoxycarbazomycinal (24), $C_{16}H_{15}NO_4$ (M$^+$ 285.1017), m.p. 221°, isolated along with carbazomycinal (74) had uv [λ_{max} 215, 227, 245, 268, 310, 382, with ε 27 000, 24 500, 13 000, 13 000, 17 000, 8500] and ir data similar to those of carbazomycinal (11) showing it to be a 1-formylcarbazole derivative. The ^{1}H-NMR spectrum showed the presence of one Ar-methyl group (δ 2.76), two methoxyl groups (δ 3.86 and 3.93) and a phenolic hydroxyl (δ 6.91). The H-5 signal (δ 7.72, 1H, d, $J = 2.4$ Hz) was *meta* coupled and H-8 was *ortho* coupled showing that the additional methoxy group was at the 6-position. Long range selective proton decoupling experiments (LSPD) in which irradiation of the H-3 signal caused the collapse of signals of both C-4b and C-6 and in which irradiation of the H-5 and H-7 signals respectively resulted in the collapse of the C-7 and C-8a signals and of the C-5 and C-8a signals showed that the alkaloid could be formulated as 6-methoxycarbazomycinal. This conclusion has been supported by ^{13}C NMR data of the compound. Nakamura *et al.* isolated both carbazomycinal and its 6-methoxy derivative and named them carbazomycin E and carbazomycin F.

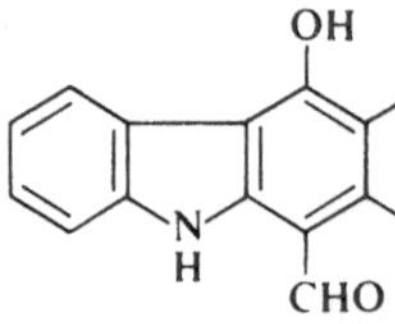

(11) 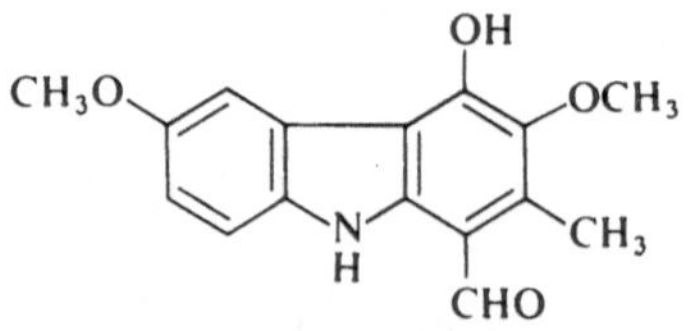   (24)

5. Carbazomycin G

Carbazomycin G (**40**), $C_{15}H_{15}NO_3$ (M$^+$ 257), m.p. 241–243° was isolated from *Streptoverticillium ehimense* (66). The uv absorption spectrum [λ_{max} 214, 253, 272, 278, 340 with ϵ 33 200, 19 800, 8600, 7400 and 5600] and the ir spectrum showed to be a carbazole derivative. The ^{1}H-NMR spectrum showed the presence of two methyl signals (δ 1.60, δ 2.01) and one methoxy signal (δ 3.7), as well as signal for H-5 (δ 8.05, 1 H, m) (deshielded by C=O at 4) H-6, H-7 and H-8 (δ 7.21–7.50). The ^{13}C-NMR spectrum showed the presence of a tertiary methyl (δ 27.9), a quaternary carbon (δ 67.3) attached to a hydroxyl as well as vinylic methyl (δ 10.10) and methoxyl (δ 59.2) groups. The ^{13}C-NMR spectrum further contained a carbonyl signal whose frequency (δ 177.5) is commensurate with 2-methoxydienone structure. The compound has an assymetric centre but is optically inactive. On the basis of physical data and the optical inactivity, carbazomycin G could be formulated as (**40**). This has been substantiated by X-ray diffraction.

6. Carbazomycin H

Carbazomycin H (**230**), $C_{16}H_{17}NO_4$ (M$^+$ 287), m.p. 228–230° was isolated from *Streptoverticillium ehimense* by Nakumura *et al.* (66). It had uv absorption and ir spectrum similar to those of carbazomycin G (**40**). The ^{1}H-NMR spectrum showed the presence of additional aromatic methoxyl group (δ 3.84) as compared with carbazomycin G (**40**). The H-5 signal was *meta* coupled (δ 7.66, d, $J = 2.44$). The diamagnetic shift of the H-5 signal by 0.57 ppm as compared with H-5 of (**40**) was attributed to the effect of the methoxyl group at C-6. Signals of the other protons of ring A and ring C were comparable with those of (**40**). The methoxy-induced shifts on the aromatic carbons in the ^{13}C-NMR spectrum of carbazomycin H also supported the placement of the additional methoxy at C-6. From all these data the structure of carbazomycin H was assigned as (**230**).

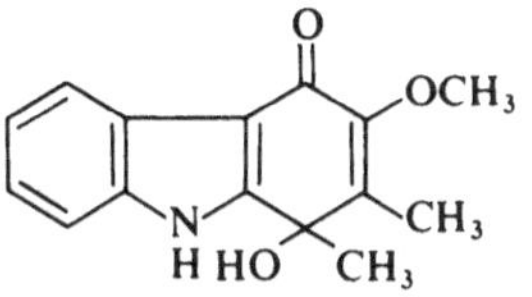

(**40**) 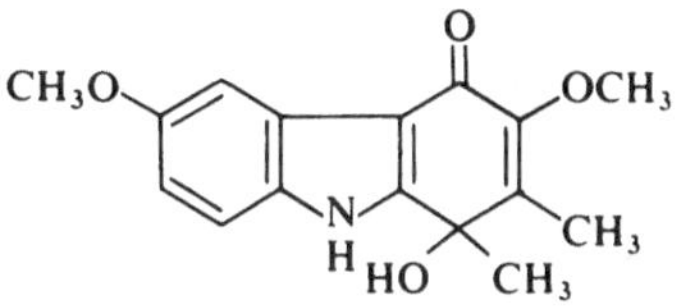   (**230**)

(ii) Kinamycins

The kinamycins constitute a group of eight antibiotic alkaloids four of which were isolated in 1970 from *Streptomyces murayamaensis* (Actinomycetes) by ITO *et al.* (*61*). They were first considered as naphthaquinone antibiotics containing an additional fused pyrrole cyclohexene ring system. As a result of recent work on the structure and biosynthesis of prekinamycin by SEATON and GOULD (*110–112*) these antibiotics can be viewed as derivatives of benzo-b-carbazole which justifies their inclusion in the carbazole group of alkaloids similar to the carbazomycins isolated from another microorganism of the Actinomycetes group. Structures of these alkaloids are based on detailed studies of kinamycin C (*54, 88, 89*).

1. Kinamycin C

Kinamycin C (**42**), $C_{24}H_{20}O_{10}N_2$ (M$^+$ 496), m.p. 150–53° $[\alpha]_D$ CHCl$_3$ − 24° was isolated from *S. murayamensis* in 1970 by ITO *et al.* The uv spectrum of kinamycin C (λ_{max} 246, 275, 370, 388 and 488 nm) and its red shift with alkali showed it to be a phenolic naphthaquinone derivative. This was also supported by the ir data. The ^{1}H-NMR spectrum showed kinamycin C to contain the following functions: one tertiary methyl (δ 1.3, s), two alcoholic acetoxy groups (δ 2.0–2.3), a hydroxyl (δ 2.57) and a hydrogen bonded phenolic hydroxyl group (δ 12.0), an isolated proton on a carbon carrying the alcoholic hydroxyl (δ 5.4, s), two protons on vicinal carbons carrying two acetoxyls (δ 5.6 and 6.2, d, $J = 7.2$ Hz each), and lastly three mutually coupled aromatic protons at δ 7.13, δ 7.5 and δ 7.6.

From the ir spectra of the diacetate (231), $C_{28}H_{24}O_{12}N_2$, m.p. 268–271° (decomp), and the O-methyl derivative (232), m.p. 140–144°, the

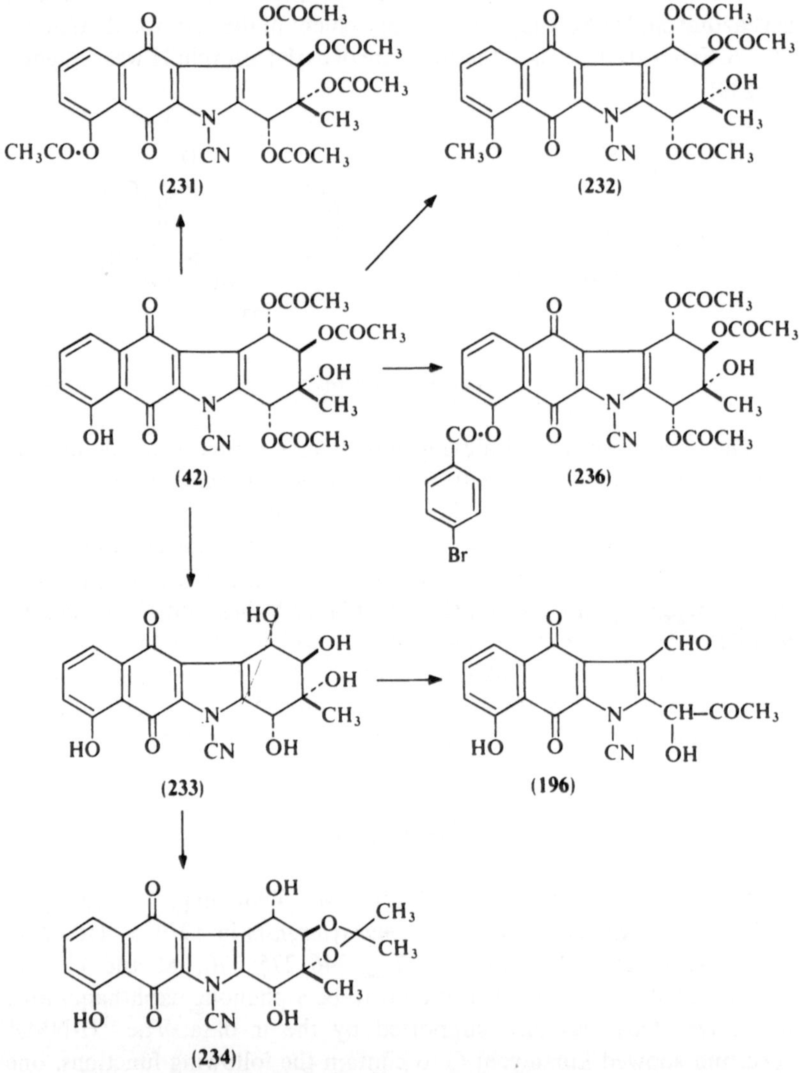

Chart 4. Reactions of kinamycin C

position of the phenolic hydroxyl *peri* to one of the quinonoid carbonyls
was inferred. From the spectral data of deacetylkinamycin (**233**),
$C_{18}H_{14}O_7N_2$, m.p. 133–36° the presence of three acetoxy groups in
kinamycin C was deduced while the ir and NMR data of deacetylkinamycin indicates the presence of an 8-hydroxynaphthaquinone fragment,
two protons on vicinal carbons carrying alcoholic hydroxyl groups, a
tertiary methyl and a nitrile or isonitrile function. Sodium periodate
oxidation of deacetylkinamycin C gave a substance, $C_{17}H_{10}O_6N_2$, m.p.
117–119° (**196**) was obtained which was an aromatic aldehydic (δ 10.52, s)
and had a methyl attached to a carbonyl group (δ 2.4, s) as well as a
benzylic proton on carbon carrying a hydroxyl group at δ 6.28 and
δ 4.28. From its ir spectrum presence of a free quinone carbonyl (v_{max}
1660 cm^{-1}) and a hydrogen-bonded carbonyl group could be deduced.
Deacetylkinamycin gave an isopropylidene derivative (**234**) with acetone,
$C_{20}H_{18}O_7N_2$, m.p. 185–188°, whose ^{1}H-NMR spectrum showed that
the tertiary hydroxyl at C-2 and the secondary hydroxyl at C-3 were
involved in the formation of the isopropylidene derivative. The structure
of kinamycin C could therefore be partially represented as (**235**).

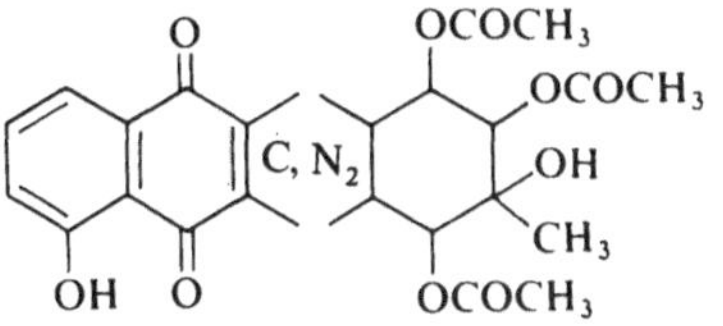

(**235**)

The nature of the CN_2 function of kinamycin C was deduced by an X-
ray analysis of the *p*-bromobenzoate derivative which established the
structure of the latter as (**236**). Hence kinamycin itself possessed formula
(**42**). Hydrolysis of deacetylkinamycin C with 10% methanolic hydrochloric acid furnished ammonia confirming the presence of the N–CN
group in the pyrrole ring.

2. Kinamycins A, B and D

Kinamycin A (**193**), $C_{24}H_{20}O_{10}N_2$ (M$^+$ 496), m.p. 139–42° (dec)
$[\alpha]_D^{20°}$ CHCl$_3$ − 60° Kinamycin-B (**195**), $C_{20}H_{16}O_8N_2$ (M$^+$ 412), m.p.
190–93° (dec) $[\alpha]_D^{25°}$ CHCl$_3$ − 48° Kinamycin D (**194**), $C_{22}H_{18}O_9N_2$
(M$^+$ 452), m.p. 170–75° (dec) $[\alpha]_D^{25°}$ CHCl$_3$ − 37°. All three compounds
have uv, ir and ^{1}H-NMR data very similar to those of kinamycin C.
Comparison of the physical properties show that all three have in

common hydroxyl, nitrile and acetoxy functions and differ only in the number of acetoxy groups.

In the three compounds both quinonoid keto groups are hydrogen bonded (v_{max} 1620 cm^{-1}) unlike the situation in kinamycin C which has both free and bonded quinonoid keto groups (v_{max} 1660 & 1625 cm^{-1}). From this one infers that they have a hydroxyl at C-4 which is bonded to the keto group. Kinamycin D on acetylation furnished diacetylkinamycin-C (M^+ m/z 580), hence kinamycin D was 4-deacetylkinamycin-C. Comparison of the chemical shifts of the tertiary methyl group in kinamycin C and kinamycins A, B, and D and their acetyl derivatives, led to the conclusion that kinamycin A has acetoxy groups at C-1, C-2 and C-3, a conclusion verified by the fact that the monoacetates of kinamycin A and kinamycin C were identical. On the other hand, tetracetylkinamycin B was identical with the diacetate of kinamycin C. From these and other data (89) the structures of kinamycin A, B, and D, can be written as (193), (195) and (194).

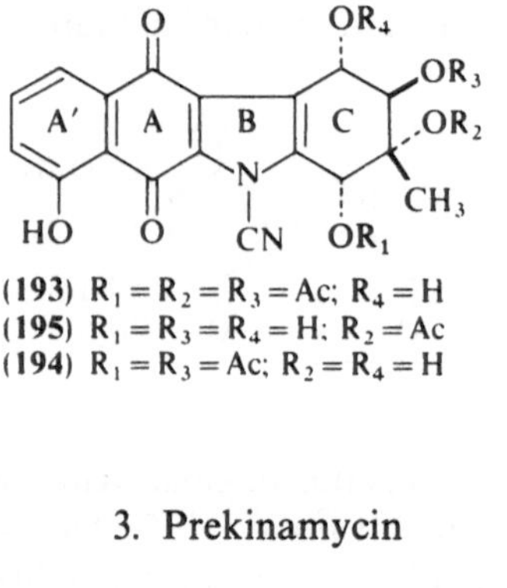

(193) $R_1 = R_2 = R_3 = Ac; R_4 = H$
(195) $R_1 = R_3 = R_4 = H; R_2 = Ac$
(194) $R_1 = R_3 = Ac; R_2 = R_4 = H$

3. Prekinamycin

Seaton and Gould isolated from *Streptomyces murayamaensis* (27, 110) prekinamycin (190), $C_{18}H_{10}N_2O_4$, m.p. 300°C (decomp). From its uv (λ_{max} 254, 288.4, 342, 574 nm, ε 5500, 21 700, 5770, 37 700) and ir spectrum it had a carbazole skeleton with a hydroxyl group chelated to a quinonoid function. The ^{1}H-NMR spectrum revealed the presence of the three adjacent H-5′, H-6′, H-7′ protons. It also showed the presence of two *meta*-coupled H-2, H-4 protons (δ 6.60 and 6.69, d, J = 1.5 Hz each) and an aromatic C-Me group (δ 2.29) besides two protons (δ 11.60 and 12.32) which were absent in its diacetate, M^+ 402 239, $C_{22}H_{16}N_2O_6$, m.p. above 300° (decomp). In the ^{13}C NMR spectrum of the diacetate the presence of two quinonoid carbonyls (δ 174.14, and 192.48), ester carbonyls (δ 170.28 and 170.64) and a cynamide function (δ 83.71) was discernible. From these data and the ^{13}C-NMR spectrum of a sample obtained biosynthetically by feeding experiments structure (190) was advanced for prekinamycin.

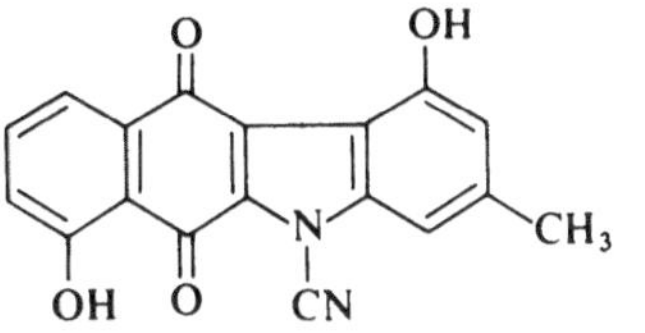

(190) 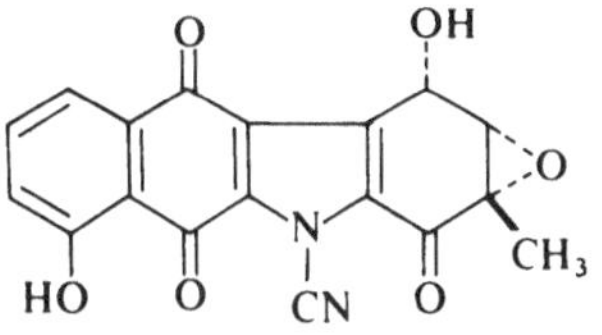   (186)

4. Ketoanhydrokinamycin

Ketoanhydrokinamycin (186), $C_{18}H_{10}N_2O_6$ (M + H$^+$ 351), m.p. 300° (decomp) was isolated by GOULD et al. (110). The uv and ir spectra reflected the closeness of the compound to the kinamycins. The ^{1}H-NMR spectrum showed the presence of a hydroxyl at δ 5.94 and a bonded phenolic hydroxyl at δ 12.06 as in the other kinamycins but lacked signals indicating acetates. It had two hydroxyls (δ 5.94 and 12.06), one of which was hydrogen bonded and the other phenolic. Its ^{13}C-NMR spectrum showed the presence of a 1,4 quinonoid chromophore and a conjugated carbonyls (δ 183.67, 180.85 and 188.62). The coupling (2 Hz) of H-3 and H-4 (δ 3.89 and 5.34) permitted the assignment of the regiochemistry shown in the formula. The resonance at δ 5.34 showed coupling to a hydroxyl at δ 5.94. A LR HETCOSY spectrum of the alkaloid showing the long range coupling between the methyl hydrogens at δ 1.53 and the carbonyl at δ 188.62 thus confirming the position of the keto group in ring C at C-1. Enhancement of the resonances at δ 3.89 and δ 5.34 by irradiation of the methyl group further confirmed the positions of H-3 and H-4 in ring C. From all these data the structure (186) was attributed to ketoanhydrokinamycin.

5. Kinamycin E

Kinamycin E (237), $C_{20}H_{16}N_2O_8$ (M$^+$ 412.0906) m.p. above 200°C (decomp) was isolated (110) from the polar fraction of the culture extract. Its uv (λ_{max} 255.6, 277.0, 295.0, 408.0 nm with log ε 21 700, 14 100, 9220, 8850) and ir spectrum suggested that it was a kinamycin analogue. The proton signal at δ 4.24 in the ^{1}H-NMR spectrum of kinamycin E could be attributed to H-3; other data showed that the acetylation pattern were similar to kinamycin D. Treatment of kinamycin D with potassium carbonate in methanol gave a small amount of kinamycin E.

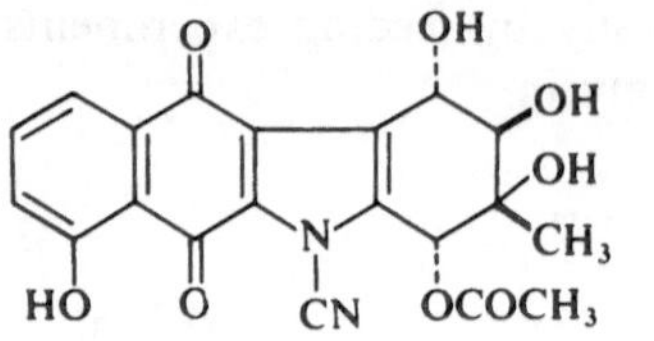

(237)

6. Deacetylkinamycin

Kinamycin F (**233**) isolated as a natural product (*110*) was found identical with deacetylkinamycin prepared by hydrolysis of kina-mycin D.

(iii) Tubingensins

1. Tubingensin A*

Tubingensin A (**20**), $C_{28}H_{35}NO$, (M^+ 401.2698), m.p. 95–98° $[\alpha]_D^{CHCl_3}$ 13.60 was isolated as a light yellow solid from the sclerotia of *Aspergillus tubingensis* (*119*). Its uv spectrum (λ_{max} 218, 239, 262, 302, 326, 340 nm with ε 14 900, 18 200, 6930, 6780, 480, 480) was similar to that of carbazole. The proton spin systems were analyzed by homonuclear decoupling experiments while the carbon assignments were made on the basis of heteronuclear shift correlation and some selective INEPT data. The isolated four spin system (δ 7.98, br d, $J = 7.8$ Hz, δ 7.18 dd, $J = 3.9$, 7.6, 7.8 Hz, overlapping multiplets at δ 7.34 and 7.34 resolved in C_6D_6) showed that ring A was unsubstituted. Singlets of H-1 and H-4 at δ 7.92 and 7.11 indicated substitutions at the C-2 and C-3 which has also been supported through INEPT correlation of H-5 with C-4a, C-4b, C-7, C-8a of H-4 with C-9a, C-4b, C-2 and C-19 and of H-1 with C-4a, C-4, C-3 and C-10. Further information was obtained from long range correlation studies. The downfield shifts of the two C-10 protons (δ 2.9 and 2.88 br. dd) were suggestive of their attachment to C-10 with a sp^2 center. The ^{1}H-NMR data showed the presence of a six carbon chain (C-20 to C-25) attached to C-13. The mass spectral peak at m/z 318 (M-83, 100%) also

---

* In consideration of the CA convention for the tricyclic carbazole system the numbering of the carbon atoms in Tubingensin A and B has been modified from that reported by the authors.

supported the presence of a six carbon chain containing a vinyl proton. Detailed INEPT studies, correlation data and biogenetic considerations led to formulation of tubingensin A as **20**.

The stereochemistry of tubingensin A follows from the NOESY correlation between H-14 and H-4. The equatorial disposition of H-14 is in accord with a *trans*-diaxial coupling between H-14 and neighbouring proton. Correlation between C-18 and C-19 showed that these methyls are gauche to each other. A cross peak correlating the axial proton at C-10 with a signal centred at δ 1.74 ppm (H-17) is consistent with *cis* orientation of H-17 and C-11 and supports the *cis* orientation of the methyl groups. The relative dispositions of C-substituent at C-12, C-17 with respect to those at C-14 and C-13 were partly verified by additional correlation.

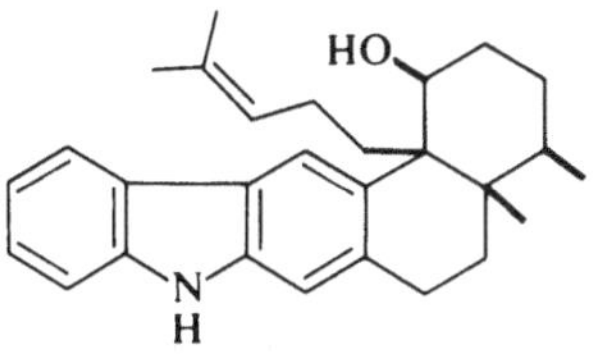

(20) 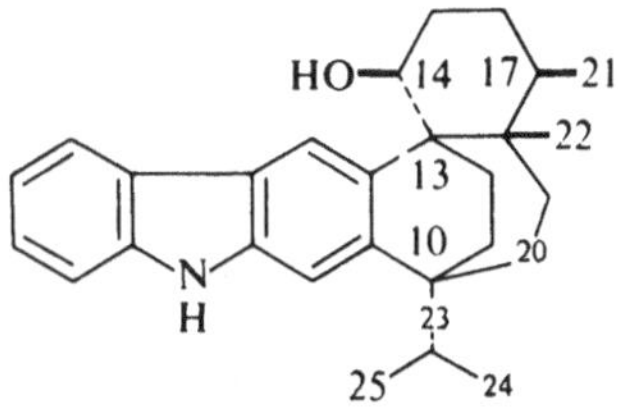   (21)

2. Tubingensin B

Tubingensin B (**21**), $C_{28}H_{35}NO$ (M$^+$ 401.2733), m.p. 152–154° $[\alpha]_D$ 6.7° (c 0.80, CHCl$_3$), uv (λ$_{max}$ 218, 237, 260, 299, 325, 338 nm, ε 17 200, 25 500, 10 100, 10 100, 2200, 6700) was isolated from *A. tubingensis* (*120*). Homonuclear decoupling and ^{1}H–^{1}H COSY experiments permitted the establishment of the spin systems. ^{13}C-NMR assignments were determined by selective INEPT experiments which afforded 2- and 3-bond CH correlations and by a heteronuclear shift correlation. These data were consistent with a 2,3-disubstituted carbazole. From the data obtained from INEPT experiments together with various other correlations structure (**21**) was attributed to tubingensin B. Its relative stereochemistry was based on the similarity of the ^{13}C-NMR spectrum to the spectra of compounds like nominine (**191**) and tubingensin A and the results of NOESY experiments. The isopropyl group at C-10 was placed *cis* to C-14 due to the spatial requirement of the ring system.

C. Alkaloids from Mammals

3-Chlorocarbazole (**212**), (M⁺ 201) was obtained from bovine urine in 1983 by LUK *et al.* (*76*). From its uv (λ_{max} 215, 223, 229, 236, 246, 260, 291, 297, 319, 330 and 343 nm; ε 2030, 2010, 2075, 2200, 1230, 1120, 600, 865, 135, 180, 150), ir and analytical data it was considered to be a chlorocarbazole derivative. This was also supported by ¹H NMR (δ 7.18–δ 7.50 (m, J) δ 7.90–δ 8.15 (m, 3). Its identity with 3-chlorocarbazole was established by direct comparison with a synthetic specimen.

(212)

References

1. AKUGIN, E., and U. PINDUR: Chemische Eigenschaften und Synthese-Verfahren von 3-Vinylindolen. J. Heterocyclic Chem. **22**, 585 (1985).

2. ALBRECHT, W.L., and R.W. FLEMING: Bis-basic Esters and Amides of Carbazoles. US. 3932456 (C.A. **84**, 150499t, 1976).

3. AL-EKABI, H., and P. DE MAYO: The Cds Photo-induced Dimerisation of N-Vinyl Carbazoles. Tetrahedron **42**, 6277 (1986).

4. ASCALONE, V., and L. DALBO: Rapid and Simple Detection of Carpofen in Plasma by HPLC with Fluorescence Detection. J. Chromatography **276**, 230 (1983).

5. BANDERANAYAKE, M., M.J. BEGELEY, B.O. BROWN, D.G. CLARKE, L. CROMBIE, and D.A. WHITING: Synthesis of Acridone and Carbazole Alkaloids Involving Pyridine Catalysed Chromene Formation: Crystal and Molecular Structure of Dibromocannabicyclol and Its Bearing on the Structures of Cyclol Alkaloids. J. Chem. Soc. Perkin I, 999 (1974).

6. BEECHAM GROUP: Antiemetics. J.P. 61 21252/1986 (C.A. **106**, 78, 761 g, 1987).

7. BERGMAN, J., and B. PELCMAN: Synthesis of Carbazoles Related to Carbazomycin, Hyellazole and Ellipticine. Tetrahedron **44**, 5215 (1988).

8. BERGMAN, J., and R. CARLSON: A Novel Synthesis of 2-Amino and 2-Hydroxy-carbazoles. Tetrahedron Letters 4051 (1978).

9. BHATTACHARYYA, L., S.K. ROY, and D.P. CHAKRABORTY: Structure of the Carbazole Alkaloid Isomurrayazoline from *Murraya koenigii* Spreng. Phytochem. **21**, 2432 (1982).

10. BHATTACHARYYA, L., S. ROY, S. CHATTERJEE, and D.P. CHAKRABORTY: Murrayazolinol—A Minor Carbazole Alkaloid from *Murraya koenigii* Spreng. J. Indian Chem. Soc. **66**, 140 (1989)

11. BHATTACHARYYA, P., and CHAKRABORTY, D.P.: Carbazole Alkaloids. In: Fortschr. Chem. Organ. Naturstoffe, Vol. 52, pp 159. (W. HERZ, H. GRISEBACH, and G.W. KIRBY, Eds.) Wien, N.Y.: Springer 1987.

12. BHATTACHARYYA, P., M. SARKAR, A.K. BISWAS, and D.P. CHAKRABORTY: Iodine

Promoted Thermal Cyclisation of Anthranilic acid. J. Indian Chem. Soc. **46**, 328 (1979).

13. BIRCH, A.J., A.J. LEIPA, and G.R. STEPHENSON: Organometallic Compounds in Organic Synthesis: Some Tricarbonyl (Cyclohexadienyl) Iron Cations and Nitrogen Containing Nucleophiles. Tetrahedron Letters 3565 (1979).

14. BISSET, N.G., A.K. CHOUDHURY, and M.D. WALKER: Occurrence of N-Cyano Alkaloids in Asian *Strychnos* Species. Phytochem. **13**, 255 (1974).

14a. CANAS-RODRIQUEZ, A., and A. MATEO-BARNARDO: Tetrahydrocarbazoles: Part-II. Tricyclic Inhibitors of Gastric acid Secretion. An. Quim. Ser. C. **81**, 254 (1985).

15. CHAKRABARTI, A., and D.P. CHAKRABORTY: Photochemical Rearrangement of Pyranocarbazole Alkaloids: Part 1. Tetrahedron Letters **29**, 6625 (1988).

16. ——: Novel Access to Bis-Carbazoles Alkaloids: Substituent Effect on the Efficiency and Regioselectivity in BF$_3$–Et$_2$O Mediated Intermolecular coupling of Pyranocarbazole Alkaloids. Tetrahedron **45**, 7007 (1989).

17. CHAKRABARTI, A., G.K. BISWAS, and D.P. CHAKRABORTY: Photo-Fries Rearrangement in N-Sulphonyl Carbazoles. Tetrahedron **45**, 5059 (1989).

18. CHAKRABORTY, D.P.: Carbazole Alkaloids. In: Fortschr. Chem. Organ. Naturstoffe, Vol. 34, pp. 299 (W. HERZ, H. GRISEBACH, and G.W. KIRBY, Eds.) Wien-New York, Springer: 1977.

19. —: Some Aspects of Carbazole Alkaloids. Planta Medica **39**, 97 (1980).

20. —: Twenty Years of Carbazole Alkaloids. Trans. Bose. Res. Inst. **47**, 49 (1984).

21. —: Some Newer Aspects of Plant Alkaloids. J. Indian Chem. Soc. **66**, 843 (1989).

22. CHAKRABORTY, D.P., B.K. BARMAN, and P.K. BOSE: On the Constitution of Murrayanine, a Carbazole Derivative Isolated from *Murraya koenigii* Spreng. Tetrahedron **21**, 681 (1965).

23. CHAKRABORTY, D.P., S. ROY, and A.K. DUTTA: Thermal Synthesis of Nor-girinimbine and Its Linear Isomer: A New Synthesis of Girinimbine. J. Indian Chem. Soc. **64**, 215 (1987).

24. CHOWDHURY, B.K., A. MUSTAFA, M. GARBA, and P. BHATTACHARYYA: Carbazole and 3-Methylcarbazole from *Glycosmis pentaphylla*. Phytochem. **26**, 2138 (1988).

25. COATS, I.H., J.A. BELL, D.C. HUMBER, and G.B. EWAN: Tetrahydrocarbazolone Derivatives, E.P. 191562 (C.A. **105**, 226579u, 1986).

26. ————: Substituted (Imidazolylmethyl)carbazolones as 5-Hydroxytryptamine Antagonists, Their Preparations and Pharmaceutical Formulation. E.P. 210840 (C.A. **106**, 15675b, 1987).

27. CONE, M.C., P.J. SEATON, K.A. HALLES, and S.J. GOULD: New Products Related to Kinamycin from *Streptomyces murrayamaensis*: Taxonomy, Production, Isolation and Biological Properties. J. Antibiotics **42**, 179 (1989).

28. DAS, B.P.: Some Investigations of the Pesticidal Properties of Carbazoles. International Pest Control **31**, 144 (1989).

29. DAS, B.P., and B. CHOWDHURY: Search for Potential Larvicides from Carbazoles. Trans. Bose Inst. **47**, 91 (1984).

30. DAS, B.: Ph.D. Thesis, Calcutta University 1982.

31. DAS, K.C., D.P. CHAKRABORTY, and P.K. BOSE: Antifungal Activity of Some Constituents of *Murraya koenigii* Spreng. Experientia **21**, 340 (1965).

32. DEAN, F.M.: Natural Coumarins. In: Fortschr. Chem. Organ. Naturstoffe, Vol. 9, pp. 225 (L. ZECHMEISTER, Ed.) Wien: Springer. 1952.

33. DEWAR, D., V. GLOVER, J. ELSWORTH, and M. SANDLER: Equal and Other compounds from Bovine Urine as Monoamine Oxidase Inhibitors. J. Neural. Trans. **65**, 147 (1986) (C.A. **105**, 72468t, 1986).

34. DOBSON, R.L.M., A.P. D SILVA, S.J. WEEKS, and V.A. FASSEL: Multidimensional Laser-based Instrument for Characterisation of Environmental Samples for Poly-cyclic Aromatic Compounds. Anal. Chem. **58**, 2129 (1986).

35. DOMANSKI, A., and J.B. KYZIOL: Synthesis of Bis-Basic Substituted 9-ethylcarb-azoles. Pol. J. Chem. **59**, 613 (1985).

36. EL-GAZZAR, M.A., A.M. EL-NAGGAR, F.S.M. AHMED, and A.M. ABDELSALEM: Synthesis and Biological Activity of Some New Substituted Aminoacyl Carbazole Derivatives. Arab. Galf. J. Sci. Res. **1**, 131 (1983) (C.A. **100**, 192230c, 1984).

37. FERRIS, R.M.: Treatment of Drug Induced Psychosis. U.S. US 4588728 (C.A. **105**, 54624m, 1986).

38. FERRIS, R.M., F.L.M. TANG, K.J. CHANG, and A. RUSSEL: Evidence that the Potential Antipsychotic Agent Rincazole (BW234 U) is Specific Competitive Antagonist of Sigma Sites in Brain. Life Science **38**, 2329 (1986).

39. FERRIS, R.M., H.L. WHITE, F.L.M. TANG, A. RUSSEL, and M. HARFENIST: Rincazole (BW234 U) Novel Antipsychotic Agent Whose Mechanism of Action Cannot be Explained by a Direct Blockade of Postsynaptic Dopaminergic Receptors in Brain. Drug. Der. Res. **9**, 171 (1987) (C.A. **106**, 27732j, 1987).

40. FILIMONOV, V.D., T.A. FILIPPOVA, V.P. LOPATINSKII, M.M. SUKHOROSLOVA, and N.V. STEPANOV: Carbon 13-NMR of N-substituted carbazoles. Transfer of Electronic Effects of Substituents Through Nitrogen to Carbazole Ring. Khim. Geterotsiki Soedin **9**, 1184 (1988) (C.A. **106**, 195772r, 1987).

41. FURUKAWA, H., C. ITO, M. YOGO, and T.S. WU: Structures of Murrayastine, Mur-rayaline and Pyrayafoline: Three New Carbazole Alkaloids from *Murraya eu-chrestifolia*. Chem. Pharm. Bull. **33**, 1320 (1985).

42. FURUKAWA, H., M. YOGO, C. ITO, T.S. WU, and C. KUOH: New Carbazoloquinones Having Dimethyl Pyran Ring System. Chem. Pharm. Bull. **33**, 1320 (1985).

43. FURUKAWA, H., T.S. WU, and C. KUOH: Dihydroxygirinimbine, a new Carbazole Alkaloid from *Murraya euchrestifolia*. Heterocycles **23**, 1391 (1985).

44. ———: Structures of Murrafoline -B and -C, New Binary Carbazole Alkaloids from *Murraya euchrestifolia*. Chem. Pharm. Bull. **33**, 12611 (1985).

45. FURUKAWA, H., T.S. WU, and T. OHTA: Bismurrayafoline-A and -B, Two Novel "Dimeric" Carbazole Alkaloids from *Murraya euchrestifolia*. Chem. Pharm. Bull. **31**, 4202 (1983).

46. FURUKAWA, H., T.S. WU, T. OHTA, and S. KUOH: Chemical Constituents of *Murraya euchrestifolia* Hayata, Structures of Novel Carbazolquinones and Other Carbazole Alkaloids. Chem. Pharm. Bull. **33**, 4132 (1985).

47. FURUSAKI, A., M. MATSUI, T. WATANABE, S. OMURA, A. NAKAGAWA, and T. HATA: The Crystal and Molecular Structure of Kinamycin C p-Bromobenzoate. Israel J. Chem. **10**, 173 (1972).

48. GALLAGHER, T., and P. MAGNUS: New Methods for Alkaloid Synthesis. Generation of Indole-2,3-Diquinomethanes as a Route to Indole Alkaloids. Tetrahedron **37**, 3889 (1981).

49. GHOSH, S., T.K. DAS, D.B. DATTA, and S. MEHTA: Studies on Enamides Part-1: Photochemical Rearrangement of N-Aroyl Carbazoles. Tetrahedron Letters **28**, 4611 (1987).

50. GLAXO GROUP: Preparation of Tetrahydrocarbazole Derivatives as Serotonin An-tagonists. Jr. Eur. Pat. Appl., E.P. 219, 193/1986 (C.A. **107**, 176032d, 1987) and previous references.

51. GRIFFITH, W.P., S.V. LEY, G.P. WHITOCOMB, and A.D. WHITE: Preparation and Use of tetra n-butylammonium Perruthenate (TBAP reagent) and Tetra-n-Propylammon-

ium Perruthenate (TPAP reagent) as New Catalytic Oxidants for Alcohols. J.C.S. Chem. Commun. 1625 (1987).

52. GRELLMANN, K.H., and U. SCHMITT: Reactivity and Decay Pathways of Photo-excited Anilino-naphthalene. J. Amer. Chem. Soc. **104**, 6267 (1982).

53. GRAEBE, C., and C. GLAZER: Ueber Carbazol. Chem. Ber. **5**, 12 (1872).

54. HATA, T., S. OMURA, Y. IWAI, A. NAKAGAWA, M. OTANI, S. ITO, and T. MATSUYA: A New Antibiotic Kinamycin. Fermentation, Isolation, Purification and Properties. J. Antibiotics **24**, 353 (1971).

55. HAUSBERG, H.H., and H. BETTCHER: Tetrahydrocarbazole Derivatives. Ger. Pat. DE 3,30094 (C.A. **101**, 210979f, 1984).

56. HUSSON, H.: Simple Indole Alkaloids Including β-Carbolines and Carbazoles In: Alkaloids, Vol. 26, pp. 1 (ARNOLD BROSSI Ed.) New York: Academic Press. 1985.

57. IGNATIADIS, I., M. KUROKI, and P.J. APPINO: Identification of Carbazole Derivatives in a Hydro Treated Gas Oil by Gas Chromatography and Gas Chromato-graphy–Mass Spectrophotometry. J. Chromatography **366**, 251 (1986).

58. ITO, C., T. WU, and H. FURUKAWA: Three New Carbazole Alkaloids from *Murraya euchrestifolia*. Chem. Pharm. Bull. **35**, 450 (1987).

59. ——: New Carbazole Alkaloids from *Murraya euchrestifolia*. Chem. Pharm. Bull. **36**, 2377 (1988).

60. ——: The Structures of Some New Carbazole Alkaloids. IUPAC Symp. New Delhi, Feb. 4–7, 1990, p. 259.

61. ITO, S., T. MATSUYA, S. OMURA, M. OTANI, A. NAKAGAWA, Y. IWAI, M. OHTANI, and T. HATA: A New Antibiotic, Kinamycin. J. Antibiotics **23**, 315 (1970).

62. JACKSON, G.D.F., and W.H.F. SASSE: Synthetical Applications of Activated Metal Catalysts. XXI. The Formation of Carbazole from Aniline and Related Compounds in Presence of Degassed Raney Nickel. Austr. J. Chem. **17**, 347 (1964).

63. JOULE, J.: Recent Advances in 9H Carbazoles. In: Advances in Heterocyclic Chemis-try, Vol. 35, pp. 83. (A. KATRITZKY Ed.) New York: Academic Press 1984.

64. KAMARUZZAMAN, S. ROY, and D.P. CHAKRABORTY: Mumpamine from *Glycosmis pentaphylla*. Phytochem. **28**, 677 (1989).

65. KANE, V., A.R. MARTIN, and P.A. PETERS: The Non-regiospecific Condensation of Citral and 2-Hydroxycarbazole. Heterocycles **16**, 1445 (1981).

66. KANEDA, M., T. NAID, T. KITAHARA, and S. NAKAMURA: Carbazomycin G and H, Novel Carbazomycin Congeners Containing a Quinol Moiety. J. Antibiotics **41**, 602 (1988).

67. KAPIL, R.S.: Carbazole Alkaloids. In: Alkaloids, Vol. 13, pp. 273. (R.H.F. MANSKE, Ed.) New York, London, Academic Press. 1971.

68. KARMAKAR, T., M. MUKHERJEE, and D.P. CHAKRABORTY: Some Urea Derivatives as Growth Inhibitors. Curr. Sci. **55**, 828 (1986).

69. KATRITZKY, A.R., G.W. REWCASTLE, and L.M.V. DE MIGUEL. Improved Syntheses of Substituted Carbazoles and Benzocarbazoles via Lithiation of the (Dialkylamino) methyl (Aminal) Derivatives. J. Org. Chem. **53**, 794 (1988).

70. KAWASAKI, T., Y. NONAKA, and M. SAKAMOTO: A New Efficient Synthesis of the 3-Methoxy Carbazole Alkaloid Hyellazole. J. Chem. Soc. Chem. Commun. 43 (1989).

71. KEDDERIS, G.L., D.E. DOUGLAS, R.N. PANDEY, and P.F. HOLLENBERG: Oxygen-18-Studies of Peroxidase Catalysed N-Methyl Carbazole. Mechanism of Carbinolamine and Carboxyaldehyde Formation. J. Biol. Chem. **261**, 15910 (1986).

72. KHAZIPOV, R. Kh., R. Kh. ALMAEV, T.V. KOROBOVA, and F.D. DAVLYATOV: Prevention of Growth of Bacteria. USSR Su 1212972 (C.A. **105** 149668 t, 1986).

73. KNÖLKER, H., and M. BAUERMEISTER: The Total Synthesis of Carbazole Antibiotic

Carbazomycin B and an Improved Route to Carbazomycin A. J. Chem. Soc. Chem. Commun. 1468 (1989).

74. KONDO, S., M. KATAYAMA, and S. MARUMO: Carbazomycinal and 6-Methoxycarbazomycinal as Aerial Mycellium Formation Inhibitory Substances of *Streptoverticillium* species. J. Antibiotics **39**, 727 (1986).

75. LONTSI, D., J.F. AYAFOR, B.L. SONDENGAM, J.D. CONOLLY, and D.S. RYCROFT: The Use of Two Dimensional Proton Coupled ^{13}C-NMR Spectrum in the Structural Elucidation of Ekebergenine, a New Carbazole Alkaloid from *Ekebergia senegenalensis* (Meliaceae). Tetrahedron Letters **26**, 4249 (1985).

76. LUK, K., L. STERN, M. WEIGEL, R.A. O'BRIEN, and N. SPIRT: Isolation and Identification of 'Diazepam Like' Compounds from Bovine Urine. J. Nat. Prod. **46**, 852 (1983).

77. LEDNICER, D., and L.A. MITCHER: In: The Organic Chemistry of Drug Synthesis, vol. 3, pp. 168. New York: John Wiley & Sons. 1984.

78. LJUNGGREN, BO, and K. LUNDBERG: *In Vivo* Phototoxicity of Nonsteroidal Antiinflammatory Drugs Evaluated by the Mousetail Technique: Photodermatology **2**, 377 (1985) (C.A. **104**, 81699 a, 1986).

79. MARTIN, T., and C.J. MOODY: A New Route to 1-oxygenated Carbazoles. Synthesis of Murrayafoline-A. Tetrahedron Letters **26**, 5841 (1985).

80. MASAKI, M.T., S. YUSHIRO, S. TSUTOMU, and N. KEIJU: Pharmacological Studies on Carpofen, a New Non-steroidal Anti-inflammatory Drug in Animals. Nippon Kagaku Zasshi **73**, 757 (1977) (C.A. **88**, 69038c, 1978).

81. MCPHAIL, A.T., T.S. WU, T. OHTA, and H. FURUKAWA: Structure of ($\pm$) Murrafoline, a Novel Biscarbazole Alkaloid from *Murraya euchrestifolia*. Tetrahedron Letters **24**, 5377 (1983).

82. MITRA, A.R.: Ph.D. Thesis, Calcutta University, 1974.

83. MOBILIO, D., C.A. DEMERSON, and L.G. HUMBER: Substituted 2,3,4,9-Tetrahydro-1-H-Carbazole-1-acetic acid and Its Use. US US4584312 (C.A. **105**, 133743c, 1986) and previous ref.

84. MOODY, C.J., and P. SHAH: Diels-Alder Reactivity of Pyrano [3,4-*b*] Indol-3-ones Part 4, Synthesis of Alkaloids Carbazomycin A, B, and Hyellazole. J.C.S. Parkin Trans 1, 2463 (1989).

85. NGADJUI, B.T., J.F. AYAFOR, B.L. SONDENGAM, and J.D. CONNOLLY: Quinolone and Carbazole Alkaloids from *Clausena anisata*. Phytochem. **28**, 1517 (1989).

86. NAID, T., T. KITAHARA, M. KANEDA, and S. NAKAMURA: Carbazomycins C, D, E, F and G. Minor Components of Carbazomycins. J. Antibiotics **40**, 157 (1987).

87. NARASHIMHAN, N.S., and S.L. KELKAR: Alkaloids of *Murraya koenigii* Part III. Structure of Currayanine and Currayangine. Ind. J. Chem. **14B**, 430 (1976).

88. OMURA, S., A. NAKAGAWA, H. YAMADA, T. HATA, A. FURUSAKI, and T. WATANABE: Structure of Kinamycin C and the Structural Relations Among Kinamycins A, B, C, D. Chem. Pharm. Bull. **19**, 2428 (1971).

89. ————————: Structures and Biological Properties of Kinamycin A, B, C, D. Chem. Pharm. Bull. **21**, 931 (1973).

90. PADWA, A., and J.A. LEE: Photochemical Transformations of 2,2-Disubstituted Chromenes: J.C.S. Chem. Commun. 795 (1972).

91. PATEL, B.P.J.: Citral Condensation: Towards Total Synthesis of Mahanimbine Isomers. Ind. J. Chem. **21B**, 612 (1982).

92. —: Cleaner Citral Condensation—Synthesis of Pyrano[2,3-a] Carbazoles. Synthetic Commun. **11**, 823 (1981).

93. —: Synthesis of Girinimbine Isomers—Pyrano [2,3-a] Carbazole Derivatives. Ind. J. Chem. **21B**, 20 (1982).

94. PATTERSON, A.M., and L.T. CAPELL: The Ring Index, p. 229, No 1675. New York: Reinhold. 1940.

95. PECCA, J.C., and S.M. ALBONICO: Trypnocides. I. Substituted 1,2,3,4-Tetrahydrocarbazoles. J. Med. Chem. **13**, 327 (1970).

96. PHARR, D.Y., P.C. UDEN, and S. SIGGIA: 3-(p-acetylphenoxy) Propyl Silane Bonded Phase for Liquid Chromatography of Basic Amines and Other Nitrogen Compounds. J. Chromat. Sci. **23**, 391 (1985).

97. PFEUFFER, L., and U. PINDUR: Diels-Alder-Reaktionen von 2'-substituierten-3-vinyl-1-H-indolen zu neuen anellierten Indol- und Carbazole-Derivaten. Helv. Chim. Acta **70**, 1419 (1987).

98. PINDUR, U.: New Diels-Alder Reactions with Vinylindoles: A Regio- and Stereocontrolled Access to Annelated Indoles and Derivatives. Heterocycles **27**, 1253 (1988).

99. PINDUR, U., and H. ERFANIAN-ARDOUST: Indole 2,3 quinodimethanes and stable cyclic Analogues for Regio- and Stereocontrolled Syntheses of [b]-Annelated Indoles. Chem. Rev. **89**, 1681 (1989).

100. PINDUR, U., and L. PFEUFFER: [4 + 2]-Cyclo-addition to 4-Demethoxycarbazomycin. Heterocycles **26**, 325 (1987).

101. ——: New Structural Aspects of 3-Vinyl-1H-Indoles for Predicting the Outcome of Diels-Alder Reactions. Monatsh. Chem. **120**, 27 (1989).

102. POINTEK, J.A., and R.Y. WANG: Acute and Subchronic Effects of Rincazole (B234U) a Potential Antipsychotic Drug on A9 and A10 Dopamine Neurons in the Rat. Life Sci. **39**, 651 (1986).

103. RAISZ, L.G., C. ALANDER, C. ONKLINKOKX, and G.A. RODAN: Effects of Thiophen-2-Carboxylic Acid and Related Compounds on Bone Resorption in Organ Culture. Calcif. Tissue Int. **37**, 556 (1985) (C.A. **103**, 206184s, 1985).

104. RICE, L.M., and K.R. SCOT: 3-Substituted 1,2,3,4-Tetrahydrocarbazoles. J. Med. Chem. **13**, 308 (1970).

105. ROY, S., A. CHAKRABARTI, and D.P. CHAKRABORTY: Lewis Acid Catalysed Aliphatic Diazo-coupling. An Access to Nitrogen Heterocycles. Proc. Convention of Chemists (Ind. Chem. Soc.) C 34 (1986).

106. ROY, S., R. GUHA, S. GHOSH, and D.P. CHAKRABORTY: Biomimetic Hydroxylation Studies on Carbazole Alkaloids. Ind. J. Chem. **21B**, 617 (1982).

107. SAKANO, K., and S. NAKAMURA: New Antibiotics Carbazomycins A and B II. Structural Elucidations. J. Antibiotics **33**, 961 (1980).

108. SAULNIER, M.G., and C.W. GRIBBLE: 4-Phenyl sulphonyl-4-H-Furo[3,4-b] Indole, a Stable Analogue of Indole-2,3-quinodimethane. Tetrahedron Letters, 5435 (1983).

109. SEARL, A.J., F.C. GEE, and R.L. WILSON: In Oxygen radical in Chem. & Biol. (Proc. 3rd Int. Conf. 1983). (W. BROS, S. MANFERED and C. TAIT, Eds.) Berlin: David de Gruyter. (C.A. **100**, 203140b, 1984).

110. SEATON, P.J., and S.J. GOULD: New Products Related to Kinamycin from *Streptomyces murayamaensis*. Structures of Pre-Kinamycin, Keto-Anhydrokinamycin, and Kinamycins E and F. J. Antibiotics **42**, 189 (1989).

111. ——: Kinamycin Biosynthesis Derivation by Excision of an Acetate Unit from a Single-chain Decaketide Intermediate. J. Amer. Chem. Soc. **109**, 5282 (1987).

112. ——: Origin of Cyanamide Carbon of Kinamycin Antibiotics. J. Amer. Chem. Soc. **110**, 5912 (1988).

113. Sengupta, S.N.: Ph.D. Thesis, Calcutta University, 1979.
114. Sexton, W.A.: In: Chemical Constitution and Biological Activity, 3rd Edn., p. 306. London: F.N. Spon Ltd. 1973.
115. Shizuki, H., M. Kato, T. Ochiai, K. Matsui, and T. Morita: The Photochemical Rearrangements of N-Acetyl Diphenylamine and N-Acetyl Carbazole. Bull Chem. Soc., Japan 43, 67 (1970).
116. Sharma, K.S., and Sarita, CMR Spectral Studies of Substituted Carbazoles. Ind. J. Chem. 27B, 402 (1988).
117. Storch, E., H. Kirchner, K. Hueller, M.G. Maritinotti, and D. Gemsa: Enhancement by Carpofen or Indomethacin of Interferon Induction by 10-Carboxymethyl-9-acridanone in Murine Cell Culture. J. Ger. Virol. 67, 1211 (1986) (C.A. 105, 35311r, 1986).
118. Stothers, J.B.: Carbon-13 NMR Spectroscopy, pp. 266. New York: Academic Press. 1972.
119. TePaske, M.R., J.B. Gloer, D.T. Wicklow, and P.F. Dowd: An Antiviral Carbazole Alkaloid from Sclerotia of *Aspergillus tubingensis*. J. Org. Chem. 54, 4743 (1989).
120. ————: The Structure of Tubingensin B: A Cytotoxic Carbazole Alkaloid from Sclerotia of *Aspergillus tubingensis*. Tetrahedron Letters 30, 5965 (1989).
121. Tursi, A., M.P. Loria, G. Specchia, and D. Cassassima: *In vitro* Studies of Anti-inflammatory Activity of Carpofen. Eur. J. Rheumatol. Inflammation 5, 488 (1982) (C.A. 98, 46621e, 1983).
122. Udenfriend, S., C.T. Clark, J. Axelrod, and B.B. Brodu: Ascorbic Acid in Aromatic Hydroxylation I, a Model System of Aromatic Hydroxylation. J. Biol. Chem. 208, 731 (1954).
123. Wood, P.L., and P.S. McQuade and S. Paul: Cyclindole and Flucindole Novel Tetrahydrocarbazoleamine Neuroleptics. Prog. Neuro-Psychopharmacol. Biol. Psychiatry 8, 773 (1984) (C.A. 102, 125458 m, 1985).
124. Yamasaki, K., M. Kaneda, K. Watanabe, Y. Ueki, K. Ishamaru, S. Nakamura, R. Nomi, N. Yoshida, and T. Nakajima: New Antibiotics Carbazomycin A and B. III. J. Antibiotics 36, 552 (1983).

(*Received August 13, 1990*)

The Bryostatins

G. R. Pettit, Cancer Research Institute and Department of Chemistry, Arizona State University, Tempe, Arizona, U.S.A.

With 7 Figures

Contents

1. Introduction

The bryostatins represent one of the medically most promising series of marine animal constituents presently known. Their discovery, chemistry and biology will be reviewed here beginning with a historical perspective.

Present molecular fossil evidence supported by geological dating strongly supports the theory that living organisms developed to the level of at least algae were prominent in oceans existing some 3.8 billion years ago (1). Many of the lower animals, especially the marine invertebrates, have been in existence for at least 1–2 billion years. Indeed ample fossil remains of hard-bodied marine animals dated to about one billion years ago have been uncovered and the animals responsible for them appear to have already reached a high degree of evolutionary development. For example, certain sponges reached their present level of evolution some 500 million years ago. Biosynthetic evolutionary processes over such an incredibly long period would favor the development of very sophisticated chemical protective mechanisms that should be very useful in various areas of modern medicine and especially in cancer chemotherapy. The fact that invertebrates do not have a thymus system responsible for immunological protection and therefore do not produce antibodies further supports this opinion (2). Obviously invertebrates must have developed mechanisms of intercellular control based on as yet unknown chemical regulation.

Invertebrates higher than the flatworm need a plasma for transporting metabolic and dissolved gas needs. Most have a pulsating organ except for members of the Annelida which have a closed vascular arrangement. In the marine invertebrates, leukocytes function by encapsulating (phagocytizing) foreign bodies and assist in wound healing. Thus phagocytosis is a primary defense system of invertebrates, presumably aided by relatively low molecular weight (non-protein) substances (3). Again present evidence suggests that antibodies are not formed in invertebrates. Thus the biosynthetic compounds utilized in such control mechanisms should be of particular importance in the discovery and development of cancer chemotherapeutic agents. Another very important consideration is the observation that cancer is very rare among marine invertebrates. With perhaps two million or more species of marine invertebrates and about one million species just in the class Insecta of the arthropods, it appears obvious that the invertebrates must have developed sophisticated defenses against neoplastic disease. Such concepts inspired this chemist in 1965 to initiate a broad and systematic program (the first such effort) to evaluate marine invertebrates (4) and arthropods (5) as sources of potentially useful anticancer drugs.

Earlier, some of the same considerations had led me in 1957 to initiate a study of amphibian venoms of the steroidal bufadienolide (6) type as potential sources of new antineoplastic substances. Eventually we found that some of the toad venom bufadienolides such as marinobufagin significantly inhibit growth of the National Cancer Institute's KB cell line derived from a human nasopharynx carcinoma and lead to a curative response with the murine Ehrlich ascites system (7). However, the therapeutic indices were unattractive for further development. So the effort was extended in 1965–66, as just noted, to encompass a geographically far-reaching area (Asia, Africa, Australia etc.) and an extensive research program to evaluate marine invertebrates and arthropods for structurally unique and useful anticancer constituents. Subsequently we isolated the first such invertebrate antineoplastic constituents (8–12). Meanwhile, our early expectations have been abundantly realized and the discovery of the bryostatins provides a splendid illustration.

Because of the rapidly increasing importance of the bryostatins as biochemical probes and in clinical trials some additional history seems appropriate. By 1968 we were able to show conclusively that some 9–10% (4) of marine invertebrates *and vertebrates* from exploratory collections displayed a confirmed level of activity against the U.S. National Cancer Institute's (NCI) murine P388 lymphocytic leukemia (PS system) or Walker carcinosarcoma 256 in the rat (13). Fortunately in 1965 two colleagues, Drs. J. HARTWELL and H. WOOD in the NCI, quickly shared my enthusiasm for exploring marine and terrestrial invertebrates for new sources of potential anticancer drugs and readily agreed to collaborate and proceed with the necessary antineoplastic evaluations. Otherwise the bryostatins would still be unknown.

To increase the geographical and species diversity, the field collections made by this investigator were supplemented by collaborations with various marine zoologists and commercial marine animal collectors. In this early period HARTWELL contacted J. RUDLOE, a very capable marine biologist, working on the North Florida coast of the Gulf of Mexico. We enlisted his aid in obtaining a selection of marine animals from that location. By this route we received in June 1968 the marine bryozoan *Bugula neritina* and quickly ascertained that extracts of this animal were exceptionally promising and capable of producing more than 100% life extension in mice bearing the PS leukemia. However, neither the Gulf of Mexico nor my subsequent, in the 1970's, Gulf of California and Gulf of Sagami (Japan) collections of *Bugula neritina* ever provided bryostatin 1 or 2. Instead these specimen collections eventually led, as summarized in the sequel, to the discovery of bryostatins 4 to 8.

By 1972 the antineoplastic activity in the Gulf of Mexico collections had disappeared. However, we were able to continue with an unidentified (at that time) Bryozoan specimen from the Gulf of California which yielded extracts and fractions with very promising antineoplastic activity. By 1976 taxonomic identification of this collection, and of other seemingly related Bryozoans from the same Gulf affording antineoplastic active extracts, showed that they were in fact all *B. neritina*. These events led me to obtain assistance from Dr. Rimmon C. Fay in supplying *B. neritina* from the U.S. coast of California, a well-known and productive habitat for this species. That recollection gave consistent PS active fractions that eventually afforded bryostatins 1–3 and revealed the Eastern Pacific Ocean *B. neritina* as the best to date source of bryostatin 1 which is about to be introduced in clinical trials. Other segments of the history and a detailed treatment of the NCI *in vitro* and *in vivo* screening results leading to selection of bryostatin 1 for clinical trial appear in an excellent recent review by Suffness and colleagues (*13*).

2. Bryozoans and *Bugula neritina* Linnaeus

The phylum Ectoprocta (Bryozoa or Polyzoa) comprises colonial filter-feeders, each member of which (polypide) is enclosed in a separate unit (zooecium). Because of their superficial appearance, marine Bryozoa are commonly known as sea-mats or false corals (*14*). Many cataclysmic events have occurred in evolution (*15*) of the phylum Bryozoa and a great number of ancient members have become extinct. The more than 4000 presently known species represent a very competitive group of animals with highly developed survival mechanisms (*15*). Most Bryozoans are marine species, but some occur in brackish and fresh water. While they live in wide areas of the world's oceans to depths over 8000 m. (*16*), they are most abundant from the warm intertidal to continental shelf depths. Perhaps due to their generally pedestrian appearance and likelihood of being mistaken for seaweeds, hydroids, or corals, these otherwise fascinating "moss animals" have remained relatively unexplored biologically and chemically (*17, 18*).

The sessile bryozoans consist of colonies of zooids formed by budding which extend into long branching chains, or disordered units. Their colonies are usually found attached to a great variety of substrates such as intertidal seaweeds, wood (dock pilings, ship hulls) and rocks. Bryozoans feed upon microorganisms using a retractable funnel prepared from their ciliated tentacles. These filter feeders prefer marine locations with rapid tidal currents. The larvae are usually very discriminating when selecting a place to settle and *Bugula flabellata* (*16*) nearly always prefers attachment to another bryozoan. As noted below this may also be a partial preference of *B. neritina* (family Bugulidae) and allow this species to complicate the biological evaluation of those organisms upon which it encroaches.

While bryozoans seem to have few predators certain species are susceptible to grazing by some molluscs and sea urchins. Dorid nudibranchs of the genus *Polycera* have a special ability to drill through the bryozoans' operculum and suck out the interior. Doubtlessly, it would be interesting and productive to explore whether the *Polycera* that ingest *Bugula neritina* in turn utilize the bryostatins for defense or other regulatory purposes.

B. neritina is known to form tuft-like colonies of up to 8 cm. in height (*16*). In our experience the usual size is about 2 cm. The zooids of this species occur on branches (stolons) in two series with individual zooids measuring $0.6–0.8 \times 0.2–0.3$ mm. In its native environment *B. neritina* is purplish-brown, turning brown to tan colored upon preservation in alcohol and subsequent drying. Fortunately *B. neritina* is very abundant in temperate ocean locations, especially those frequented by ships as this is a most serious ship-fouling species. Additionally *B. neritina* has been observed in Japan to overwinter and produce two generations per year from the surviving bases (*16*). Because of its economic importance as a fouling organism methods have been developed for its production by laboratory culture (*13*) and eventually the bryostatins will be produced by large scale mariculture methods.

3. Isolation and Structure Elucidation of the Bryostatins

3.1. Bryostatins 1–3

When we found (*4*) that extracts of the marine bryozoan *Bugula neritina* L. exhibited exceptional antineoplastic activity (100% life extension) against the PS lymphocytic leukemia we proceeded with an extensive investigation of this animal and 14 years later reported (*17*) the isolation and x-ray crystal structure of bryostatin 1 (**1**), the first member of a very potent (low dose) class of structurally new antineoplastic substances. The final isolation procedure utilizing PS guided separation of 500 kg of *Bugula neritina* collected in the Eastern Pacific Ocean (California coast) afforded 0.12 g (2.4×10^{-5}% yield) of bryostatin 1 (**1**, NSC 339555) as colorless crystals (from methylene chloride-methanol) melting at 230–235° (*17*). A year later (1983) we described the isolation (6.3×10^{-5}% yield) and structural elucidation of bryostatin 2 (**2**, NSC 339554) (*18*). The structure was assigned employing high resolution ^{1}H- and ^{13}C-NMR spectroscopy and mass spectral interpretation combined with results of selective micro-scale acetylation–deacetylation experi-

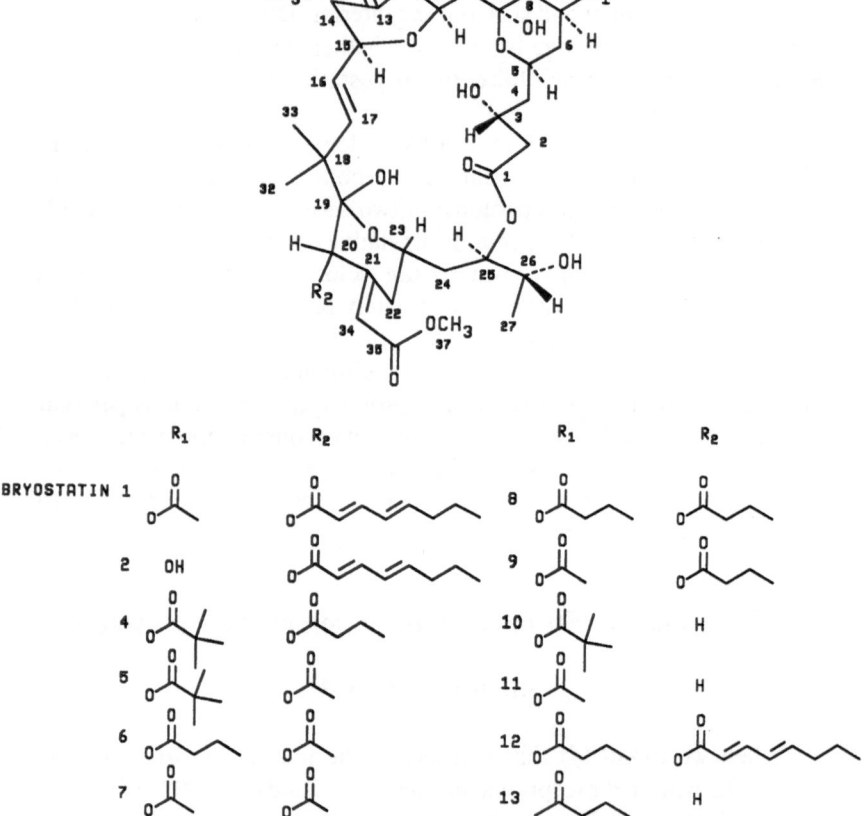

Fig. 1. Structures of bryostatins 1, 2, 4–13 and numbering system

ments. Details of the NMR high field (400 MHz) studies appear in the Spectral Characterization section.

More detailed bioassay guided separation (19) of *Bugula neritina* (Eastern Pacific) PS active fractions led to discovery of bryostatin 3 (3, NSC 350859) in $1.6 \times 10^{-7}\%$ yield (amorphous solid). Analysis of the FAB mass spectra of bryostatins 1 (1) and 3 (3) indicated a net loss of 16 mass units for bryostatin 3. Furthermore, the FAB fragmentation pattern indicating loss of the acetate and octa-2,4-dienoate side chains of bryostatin 1 was repeated in the spectrum of bryostatin 3. The similarity of optical rotations and ultraviolet spectra exhibited by bryostatins 1–3 also suggested that the new macrocyclic lactone (3) resembled its companions. The ^{13}C-NMR spectra led to the same overall conclusion and suggested that bryostatins 1–3 contained the same bryopyran ring system, but some signal shifts (1–3 ppm) and other differences were noted. Signals previously assigned to carbons C-21 and C-47 of bryostatins 1 and 2 were missing, and that due to C-35 appeared to undergo a downfield shift from 166.8 to 171.9 ppm. The ^{13}C-NMR spectrum of lactone 2 also contained new signals at 41.82, 68.30, 114.12, and 166.73 ppm believed to correspond to C-21, C-34, C-22, and C-23, respectively. However, the fundamental ring system we had designated (19) bryopyran (4) was still represented in bryostatin 3. Except for a profound change in the bryopyran C-ring, the remainder of bryostatin 3 was found to be essentially identical with bryostatin 1 (1). A shift

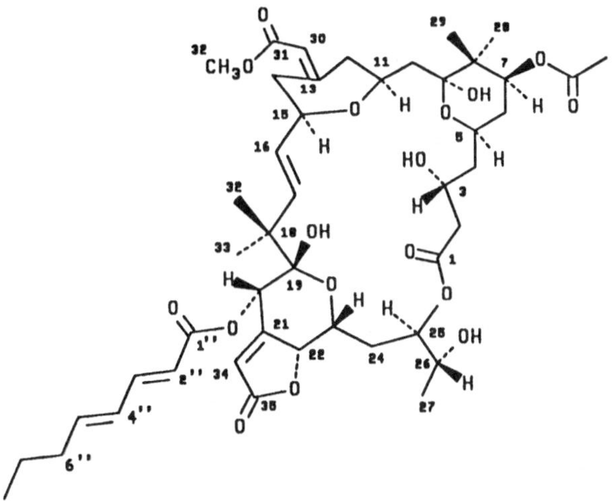

2, BRYOSTATIN 3

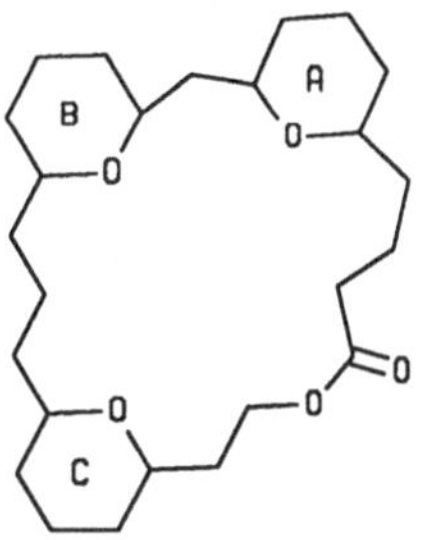

BRYOPYRAN

Fig. 2. Structure of bryostatin 3 and the bryopyran system

(C-21, 34–C-22, 23) in the exocyclic ring C olefin of bryostatin 1 to a dihydropyran system and hydroxylation at C-22 followed by lactonization with the C-35 carbonyl of the methyl ester side chain provided a satisfactory explanation of the bryostatin 3 to bryostatin 1 relationship. The preceding evidence combined with the unequivocal x-ray crystal structure determination of bryostatin 1 (1) allowed structure 3 to be assigned to bryostatin 3.

3.2. Bryostatin 4

Collections of *B. neritina* from the Gulf of Mexico (USA), Gulf of California (Mexico), and Gulf of Sagami (Japan) were under study (*20*) in parallel with the above research. *B. neritina* from these geographically diverse marine areas were found to contain a new and very potent cell growth inhibitory substance designated bryostatin 4 (4, NSC 362621) that differed markedly with respect to side chain esterification from the California series (1, 2). Isolation of bryostatin 4 was accomplished using the methylene chloride fraction from a methylene chloride–methanol–water extraction of *B. neritina* (50 kg damp wt., Gulf of Mexico) which was partitioned between 9:1 methanol–water and hexane. The methanol–water phase was diluted to 4:1 methanol–water and extracted with methylene chloride. Overall separations were guided by bioassay (PS system), and a variety of isolation procedures were investigated. The following techniques proved most effective. The methylene chloride fraction was subjected to a series of gel permeation (Sephadex LH-20, 2:3 and 2:1 methylene chloride–methanol), silica gel (hexane–acetone, hexane–ethyl acetate), and reversed phase (C-18, methanol–water) HPL chromatographic procedures. Bryostatin 4 (4) was obtained (44.5 mg, $8.9 \times 10^{-5}\%$ yield) from methylene chloride–methanol as an amorphous powder.

Discovery of bryostatin 4 (4) as a prominent antineoplastic constituent of *Bugula neritina* found in such diverse geographical areas as the Gulfs of Mexico (U.S.) and Sagami (Japan) suggested that this fascinating macrocyclic lactone system might well be a biosynthetic product of the animal rather than be derived from a dietary source (bacteria, phytoplankton or other microorganisms). But a definitive conclusion regarding this important biochemical question will require a careful chemical examination of the microorganisms ingested by *Bugula neritina* and/or (^{14}C)-acetate biosynthetic feeding experiments. Such experiments are now underway in other laboratories.

3.3. Bryostatins 5–7

Careful separation (PS bioassay) of bryostatin 4 companion fractions from the American (Gulf of Mexico), Mexican (Gulf of California), and Japanese (Gulf of Sagami) specimens of *Bugula neritina* led (*21*) to bryostatins 5–7 (**5–7**). A total of 50 kg (damp wt.) of *B. neritina* from the Northeastern Gulf of Mexico gave bryostatins 5–7 in 14.1 mg. (needles, mp 169–172°, NSC 362616), 61.9 mg. (needles, mp 172–175°, NSC 362617), and 31.2 mg. (needles, mp 176–179°, NSC 362619) yields respectively. Structural assignments (*21*) for each of these potentially important substances were derived from detailed interpretation of the 400 MHz ^{1}H-NMR and solution phase secondary ion mass spectral measurements, combined with the results of selective acid-catalyzed (1% hydrochloric acid in aqueous methanol) hydrolyses (*21*) of the C-7 and C-20 ester substituents.

From evidence now in hand, *Bugula neritina* indigenous to the U.S. West Coast seems to have a special ability for synthesis of the C-20 (*E, E*)-octa-2,4-dienoate ester. While no evidence for even trace amounts of bryostatins 1–3 could be detected in the various Gulf specimens of *Bugula neritina*, we were able to find barely detectable (by thin layer chromatography) amounts of bryostatins 4–7 in the Eastern Pacific *Bugula neritina*. The trace amounts (tlc only) of bryostatins 4–7 came from about 500 kg (wet wt.) of California *B. neritina*.

3.4. Bryostatin 8

The discovery of bryostatin 8 (**8**, amorphous, mp 170–173°, NSC 369872) resulted from a long term study of another bryozoan yielding antineoplastic constituents (*22*). A 1968 collection of *Amathia convoluta* (*4*) from the Northeastern Gulf of Mexico was found to yield extracts which more than doubled the life span of animals bearing the PS lymphocytic leukemia. Indeed, this was an exceptionally promising lead and was vigorously pursued until brought to fruition in 1983. While biologically and chemically fascinating, the physical appearance of *Amathia convoluta* is deceptively mundane. Perhaps this accounts in part for the absence of any prior chemical study and biological investigation beyond distribution and taxonomy. *A. convoluta* occurs in small shrub-like clusters from 1 to 3 cm tall, where each individual animal of the greyish-brown colony approximates 0.1 mm in size. In general, *A. convoluta* resembles the bryozoan *Bugula neritina* L. and we have found both to occur together in certain areas of the Northeastern Gulf of Mexico. That observation became very important. Application of the experimental techniques developed for discovery of bryostatins 1–4 provided solutions to the challenging research problems offered by 100 kg (damp wt.) of *Amathia convoluta* (*22*). Bryostatins 4 (**4**, 6.0 mg) and 6 (**6**, 5.7 mg) were isolated accompanied by bryostatin 5 (**5**, 3.1 mg) and a new bryostatin designated 8 (**8**, 4.2 mg). The combined total yields of bryostatins 4 and 6 amounted to 7.6×10^{-6} and $6.3 \times 10^{-6}\%$, bryostatin 5, $3.1 \times 10^{-6}\%$ and bryostatin 8, $4.2 \times 10^{-6}\%$.

Once structures for the principal antineoplastic constituents of *A. convoluta* were established, a possible relationship between this animal and the related *B. neritina* came under scrutiny. Voucher specimens from all prior collections of *A. convoluta* were closely examined. Amounts ranging from 0.2% to approximately 3% dry weight of *B. neritina* were found growing on *A. convoluta* in a parasitic or epiphytic-like manner. A 1981 recollection was found to have approx. 2.5% by weight of attached *B. neritina*. From our experience with the isolation and characterization of bryostatins 4–7 from *B. neritina* collected in the same general area as *A. convoluta*, the yields of bryostatins 4–6 seemed to be two to four times greater than would be expected from *A. convoluta* containing some 2.5% of *B. neritina*. But this proportion was too close to permit the safe conclusion that they were produced by *A. convoluta*. On the other hand bryostatin 8 (**8**) was isolated from *A. convoluta* in a quantity 50 times greater than would be expected from the companion *B. neritina*. Thus, bryostatin 8 seemed to be a genuine constituent of *A. convoluta* and/or both animals have a relationship where the bryostatins may be transferred and concentrated by the *A. convoluta*. Alternatively, if the real source of the bryostatins resides in a common food source such as a dinoflagellate or other microorganism, these relationships may be even more complex.

Further evidence for the ability of *B. neritina* to invade other marine organisms was accumulated from our study of a marine tunicate. By early 1973, we had already collected the Gulf of California ascidian *Aplidium californicum* Ritter and Forsyth for antineoplastic evaluation. *A. californicum* is found as yellow-to-orange pulpy masses (sea pork) growing on the submerged portions of docks and other hard substrates exposed to strong tidal currents. Extracts of this animal reached the confirmed active stage during 1975 in the NCI PS leukemia affording a 68% increase in life span at a dose of 60 mg/kg. Four recollections of the tunicate made from 1976–1982 gave quite variable biological results ranging from loss of activity to the 1978 sample with good results (approximately that of the original specimens, *23*). Extracts from the

1978 recollection (34 kg wet wt.) gave a 45% life extension increase against the PS leukemia and were considered sufficiently active to complete the study of *A. californicum*. A combination of results from the previous investigation with some of the best experimental techniques (*22*) we had developed to that date for solving such challenging problems led to isolation of bryostatins 4 (0.9 mg) and 5 (0.5 mg) Careful examination of the museum specimens preserved from the *A. californicum* recollections revealed that all contained small quantities of interstitial *B. neritina* undetected during initial cleaning and processing of the original animal collection. Because ascidians continue to secrete an external connective tissue from living cells passing out through the epidermis, it is not clear whether the *B. neritina* had invaded *Aplidium* or was encased in the growing skin (a tunic or test). The arthropod *Polycheria osborni* invades *A. californicum* by burrowing into the tests (*24*), but any invasion route by *B. neritina* would probably involve passage with the feeding water from pharynx to atrium. Whatever the mechanism, *Aplidium* may enjoy its relationship with *Bugula* for defense purposes. Indeed, the luxuriant growth of *A. californicum* at the Bahia Kino, Sonora, collection site suggests a symbiotic relationship with *B. neritina*.

As emphasized above in the summary of *A. convoluta* research, *B. neritina* contains very potent antineoplastic constituents with the capacity to intrude upon or otherwise become associated with certain other marine organisms. Therefore, the possibility of *Bugula* contamination should be considered prior to selecting a new marine animal for detailed chemical study directed at isolation of possible antineoplastic constituents.

Aside from the capricious result of the *A. californicum* research the tunicates in general represent a productive source of new antineoplastic substances (for leading references consult (*23*)). Indeed, during our initial (1965–1968) worldwide evaluation of marine animals as sources of new anticancer substances, we uncovered the first tunicates (e.g., *Styela plicata*) with such constituents (*4*). On our Institute's 1976 expedition to the coast of Honduras, we located a very promising [NCI confirmed active, T/C 173 (9.37 mg/kg)] tunicate of the *Trididemnum* genus that was later found independently by A. J. WEINHEIMER and colleagues to yield potent antineoplastic cyclic depsipeptides (the didemnins).

3.5. Bryostatin 9

The accelerating interest arising from the remarkable biological properties displayed by bryostatins 1 and 2 encouraged us to look for further members of this series. When a 1982 Gulf of Mexico (Florida) recollection of *B. neritina* (50 kg wet wt.) was examined using PS bioassay guided separation of minor antineoplastic constituents bryostatin 9 (**9**, needles mp 159–162°) was isolated (13.5 mg) in $2.7 \times 10^{-5}\%$ yield (*25*). Bryostatin 9 was also detected in admixture with bryostatin 11 in fractions from a Gulf of California (Sonora, Mexico) recollection. The structure was established as bryopyran **9** by a detailed analysis of the 400 MHz ^{1}H-NMR and solution phase secondary ion mass (SP–SIMS, *26*) spectral data. The 400 MHz ^{1}H-nmr spectrum of bryostatin 9 was found to be very similar to that of bryostatin 6. Also, the mass spectrum of bryostatin 9 gave ions at m/z 875 $[M + Na]^+$, $[M^+$, 852 for $C_{43}H_{64}O_{17}]$, 815 $[M + Na-60]^+$, and 787 $[M + Na-88]^+$ as previously found for bryostatin 6 ($C_{43}H_{64}O_{17}$, butyrate, and acetate esters at C-7 and C-20) and again suggested the presence of both acetate and butyrate esters. Thin layer chromatographic analyses of bryostatins 6 and 9 showed analogous, but not identical, mobilities. The corresponding acetates, prepared by acetylation with acetic anhydride–pyridine, were also different. Therefore, the acetate and butyrate substituents of bryostatin 9 appeared to be reversed at positions 7 and 20 from those of bryostatin 6.

To verify the tentative structural assignment, bryostatin 9 was treated with 1% hydrochloric acid in methanol at ambient temperature for 3 days. The crude product was purified using hplc (reversed phase C-18 chromatography) to yield (1 mg → 0.65 mg) as major product the C-20 butyrate minus the C-7 acetyl group. The hydrolysis product was found to be identical with the principal acid-catalyzed cleavage product from bryostatin 8. From this evidence combined with a spectral analysis, the 7-hydroxy-20-butyrate structure was assigned to the bryostatin 9 hydrolysis product and thereby structure **9** to bryostatin 9.

3.6. Bryostatins 10 and 11

Determined investigation of biologically active (PS) trace fractions from Gulf of California and Gulf of Mexico collections of *Bugula neritina* led to the 20-desoxybryostatins 10 (**10**) and 11 (**11**) which represented a new class (*28*). The biological activity and physical properties of bryostatins 10 and 11 (plates, mp 161–164° and needles, mp 171–173°, NSC 352697 and 606642, respectively) closely approximated those of bryostatin 4. While these properties greatly complicated their isolation and purification, they did prove useful for simplifying initial characterization. Since bryostatin 10 closely resembles bryostatin 4 and was obtained in somewhat higher yield (6.7×10^{-5} *vs* 1.6×10^{-5}%) the 33.4 mg obtained from 50 kg (wet wt.) of the Gulf of Mexico (Florida) *B. neritina* was subjected to structure determination. Information derived from that study assisted in assigning structure **11** to the even rarer bryostatin 11 (8.5 mg, 2×10^{-7}% yield) from the same source. Definitive structural assignments for bryostatins 10 and 11 were based on analysis of ^{13}C-NMR and 400 MHz ^{1}H-NMR spectral data combined with results of selective acetylation, dehydration, and oxidation experiments.

The molecular formula $C_{42}H_{64}O_{15}$ for bryostatin 10 was derived from the solution-phase secondary ion mass spectrometry (SP-SIMS) procedure (*26, 27*) that we developed employing sodium iodide or silver tetrafluoroborate in sulfolane solution. By this means large and easily interpreted molecular ion and fragment ion complexes were formed. The SP-SIMS fragmentation loss of pivalate was observed as with bryostatin 4 but not the loss of butyrate. These observations were the first clue to the possible existence of a new type of bryostatin bearing a methylene group at C-20. Further evidence for that assumption was obtained by examining the ^{13}C-NMR and ^{1}H-NMR spectra obtained from bryostatin 10. The carbon resonance position for carbons 1-19, 22-33, 35-37, and 1'-5' were almost identical. Most obvious was the absence of the butyrate side chain and the upfield shift of the C-20 signal from $\delta 79.13$ in bryostatin 4 to $\delta 36.17$ in bryostatin 10. The adjacent olefin at C-21 → C-34 carbon resonances were shifted from $\delta 151.88$ to 156.98 and from 119.68 to 115.75 respectively. Confirmation for the absence of oxygen at C-20 was obtained by comparing the high-field (400 MHz) ^{1}H-NMR data derived from bryostatins 4 and 10. A series of chemical experiments summarized in Section 5 completed the structural assignment for bryostatin 10 and an analogous approach was used to assign the structure of bryostatin 11. Since bryostatin 4 also occurs in *B. neritina* from the Gulf of Sagami (Japan) we anticipate that bryostatins 10 and 11 will eventually be located in this and other sources.

3.7. Bryostatins 12 and 13

In order to isolate and structurally identify the remaining bryostatins earlier detected only in microgram or less quantities and to obtain enough bryostatin 1 for homonuclear correlation (COSY), two dimensional carbon–proton chemical shift correlations (^{1}H- ^{13}C 2D), 2D-J-resolved experiments, proton–proton differential nuclear Overhauser enhancement (NOEDS, or ^{1}H-[^{1}H] NOE) experiments, and additional biological evaluation, a 4000-L recollection (1981) of *Bugula neritina*

($\sim$ 1000 kg damp wt.) from the Eastern Pacific Ocean (California) was extracted (over 30 months) with 2-propanol and the product subjected to careful separation guided by bioassay employing the PS leukemia. The initial large-scale separation procedure was based on our earlier techniques (29–31) except for introduction of a process-scale high-pressure liquid chromatography (HPLC) step following partition of the first methylene chloride fraction between methanol–water (9:1) and hexane. By this means the methanol–water fraction (1.8 kg) was separated into 26 major fractions which were subjected to a series of chromatographic separations. The bryostatin-enriched fractions yielded as major components bryostatin 1 (1) and bryostatin 2 (2). The residual fractions from separation of bryostatins 1 and 2 were further fractionated by reversed-phase HPLC (C-18, methanol–water gradient). In this manner bryostatins 3 (1.6 mg), 8 (13.2 mg) and 9 (16.4 mg) were obtained. The latter two bryostatins had not previously been located in the California *B. neritina.* More importantly, two new substances designated bryostatins 12 (**12**, 3.7 mg, 4.0×10^{-7}% yield) and 13 (**13**, 0.7 mg, 7.0×10^{-8}% yield) were discovered (32). A detailed series of ^{1}H–^{1}H COSY, 2D-J-resolved, ^{1}H- ^{13}C 2D-shift correlated, and ^{1}H-[^{1}H] NOE difference ^{1}H-NMR experiments (see Section 7) combined with ^{13}C-NMR and SP-SIMS studies were employed to elucidate the structures of bryostatins 12 and 13.

3.8. Bryostatins A and B

Almost from the start of our evaluation of marine animals for antineoplastic constituents we have examined a broad selection of sponges from far-reaching geographical locations. As part of this effort, specimens of the Gulf of California yellow sponge *Lissodendoryx isodictyalis* were collected in early 1972 (33). Later a 2-propanol extract reached the NCI confirmed active level by leading to 45% life extension (at 12.5 mg/kg) against the PS leukemia. The antineoplastic activity displayed by the initial Mexican collection and subsequent recollections from 1976–1982 was surprising. Six collections (1968–1974) of this yellow (interior) and very fibrous sponge from the coast of Florida and the Caribbean did not give PS active extracts (13).

Because only very limited quantities of Gulf of California *L. isodictyalis* were available from recollections made at various times from 1976–1978 and because of the substantial challenges involved in isolating what appeared to be a series of complex antitumor constituents present in only trace amounts, all efforts to solve this problem until 1982 were unrewarding. During a 1976 expedition we explored some 1500 miles of the eastern Gulf of California and Pacific Ocean coasts of Mexico and found the yellow sponge in only one location, the original collection site in Bahia de Kino, Sonora. The sponge was originally found under an abandoned dredge damaged by a storm and has grown from a few kg. when first sampled in 1973 to considerably more than 100 kg. when a 108 kg. recollection was made in early 1981. In the intervening period, almost the same amount had been removed for preliminary studies. In that habitat *L. isodictyalis* obviously flourished; this may have been due to a symbiotic relationship with *B. neritina* and its potent cell growth inhibitory constituents.

A challenging series of bioassay guided separation procedures (see Section 4) led to a series of active fractions (*33*). The reddish-purple color exhibited by side products resulting from treating the antineoplastic active compounds on silica gel thin layer plates with an anisaldehyde–acetic acid–sulfuric acid mixture suggested the presence of bryostatins. From this clue four of the six antineoplastic constituents were identified as bryostatins 4–6 and 8 obtained in respectively 3.0×10^{-6}, 2.0×10^{-6}, 7.4×10^{-7} and $1.0 \times 10^{-6}\%$ yields. The remaining two active compounds seemed to be new bryostatins bearing one oxygen atom less than those previously discovered. With this unusual oxygen content they appeared to be enzymatic transformation products and were not assigned a numerical position in the sequence. Instead they were designated bryostatins A and B (8.3×10^{-7} and $6.48 \times 10^{-7}\%$ yields). Due to the micro quantities (900 µg of A and 700 µg of B) isolated, the principal structural information was derived from solution phase-SIMS and high resolution ^{1}H-NMR spectra. Unfortunately we were unable to obtain enough of A and B for a ^{13}C-NMR analysis that would have led to more definite structural assignments.

When museum specimens of the sponge were carefully examined we found *B. neritina* intrusions to the extent of some 2–5% by weight. The 1981 recollection which had undergone a thorough exterior cleaning prior to extraction contained an interior amount closer to 2%. Doubtlessly, the bryostatins obtained from the yellow sponge extracts were derived from *B. neritina* and the yields (*cf.*, Table 1) of bryostatins 4–6 were as would be expected if the material contained only a few percent by weight of the bryozoan. But on the assumption of even up to 5% invasion by *B. neritina* the yield of bryostatin 8 was considerably higher ($\sim 500\%$) than would be expected. As expected, a 2-propanol extract of *B. neritina* removed from the sponge led to 54% (at 6.2 mg/kg) life extension in the PS system.

Table 1 contains a summary of bryostatin yields from Bahia de Kino, Sonora and Gulf of Mexico *B. neritina* compared with those obtained

Table 1. *Comparison of Bryostatin Content in Four Marine Animals*

Marine animal	Amount (wet wt kg)	Bryostatins (mg)							Location (date)
		4	5	6	7	8	A	B	
Bugula neritina	12	12.3	3.0	2.8	0.6	0.5	...	...	Gulf of California Mexico (1982)
	50	44.5	14.1	61.9	31.2	1.5	...	...	Gulf of Mexico, Florida (1982)
Lissodendoryx isodictyalis	33.7	1.0	1.2	...	...	0.5	...	...	Gulf of California (1978)
Aplidium californicum	108	3.2	2.5	0.8	...	1.1	0.9	0.7	Gulf of California (1981)
	34	0.9	0.5	...	...	...	...	...	Gulf of California (1978)
Amathia convoluta	245	7.6	3.1	6.3	...	4.2	...	...	Gulf of Mexico (1981) Florida

from intrusions of this animal into *L. isodictyalis* (yellow sponge), the Bahia de Kino tunicate *Aplidium californicum* and the Florida bryozoan *Amathia convoluta*. The finding that bryostatin 7 was missing, that bryostatin 8 was obtained in much larger quantity than expected and the discovery of new members A and B suggests that some concentration and/or enzymatic transformation of these antineoplastic constituents occurs in the sponge. Indeed the robust development of *L. isodictyalis* over the near decade of study suggested a very fruitful symbiotic relationship with *B. neritina*. Such adventures by this exceptionally resourceful and adaptable bryozoan bearing highly active antineoplastic constituents demand caution when new marine animal candidates are evaluated for more detailed biological and chemical examination.

Apparently, *Bugula neritina* in the Gulfs of California, Mexico and Sagami contains the biosynthetic capability and/or the necessary exogenous microorganism(s) to produce or concentrate bryostatins with simple aliphatic ester substituents, particularly butyl, at the C-7 and C-20 positions. By contrast, Eastern Pacific Ocean *Bugula neritina* is almost devoid of that capability and instead is best at producing or concentrating the C-20-octadienoate ester type bryostatins. Discovery of the 20-deoxybryostatin **13** in $7 \times 10^{-8}\%$ yield suggests that the 20-deoxybryostatins will generally be quite rare in this bryozoan. The unequivocal structures assigned bryostatins 1–13 combined with their remarkable biological properties now provide the necessary foundation for molecular modeling and extended structure–activity investigations. Furthermore, discovery of the bryostatins amply substantiates our early expectations (*4*) that the marine animals will prove to be an exceptionally productive and valuable source of new cancer chemotherapeutic and antiviral drugs as well as providing especially useful biochemical probes.

4. Bioassay Guided Isolation

As screening capacity in the U.S. National Cancer Institute's P388 lymphocytic leukemia (PS) test increased in the 1965–1968 period it began to have a very beneficial impact on the success of our research directed at the discovery of marine animal antineoplastic constituents. The early (*6*) and more recent (*13*) evolutionary direction of the NCI antineoplastic and cytostatic evaluation systems has been reviewed. By 1968 extracts of *B. neritina* reached the confirmed active level and consistently allowed over a 100% increase in life span. Bioassay guided isolation was begun with the PS *in vivo* system and supplemented from 1975 on with the PS *in vitro* cell line. The *in vivo* and *in vitro* activity correlated well and increasing use was made of the faster cell line technique.

Bioassay directed isolation of bryostatins 4–6 and 8 from the Gulf of Mexico bryozoan *Amathia convoluta* proved to be very challenging (*22*). Indeed solution of this research objective required assaying many separation techniques and thousands of fractions. Schemes 1a–c summarize the most effective procedure (*22*) for isolating trace quantities of the bryostatins from such a relatively large initial biomass (see also the procedures in references *23* and *32*). Our pilot plant scale method for isolating bryostatins 1 and 2 has been summarized in references *13* and *32*. Here should be noted that the bryostatins can be recognized on silica gel thin layer plates by the reddish-purple color produced upon development with an acetic acid–anisaldehyde–sulfuric acid spray reagent.

Chemical investigation of *A. convoluta* began in 1970 with spring and fall recollections totalling some 50 kg. During the next six years various methods for extraction of these collections and separation of active constituents were investigated to no avail. However, encouraging progress was noted when the aqueous 2-propanol extract was partitioned between methylene chloride and water followed by successive partitioning of the methylene chloride fraction between $(9:1 \rightarrow 4:1 \rightarrow 3:2)$ methanol–water and hexane $\rightarrow$ carbon tetrachloride $\rightarrow$ methylene chloride. By this means the PS *in vivo* activity was concentrated in the carbon tetrachloride fraction.

The above very useful technique derived from studies more than 30 years ago using methanol–water as upper layer and chloroform as lower layer for extraction of lipids from biological materials (*29*). By 1959 such experiments were developed into a practical technique by BLIGH and DYER for extracting marine vertebrate lipids (*30*). Later a $9:1 \rightarrow 4:1 \rightarrow 3:2$ methanol–water with ligroin $\rightarrow$ carbon tetrachloride $\rightarrow$ chloroform variation was productively used by KUPCHAN and colleagues and by our group for preliminary investigation of higher plants containing antineoplastic components. By 1975, because of safety considerations and side-reactions from carbene, hydrogen chloride and phosgene formation, chloroform was replaced by methylene chloride, a substitution which has proved to be even more useful in terms of concentrating biologically active constituents. Further improvement in the overall preliminary separation procedure was achieved by development of a very simple and effective extraction procedure for animal and plant materials employing 1:1 methylene chloride–methanol. After extraction the solvent mixture was diluted with sufficient water to cause phase separation and the methylene chloride fractions were collected. Sufficient methanol and methylene chloride was added to the aqueous methanol mixture to

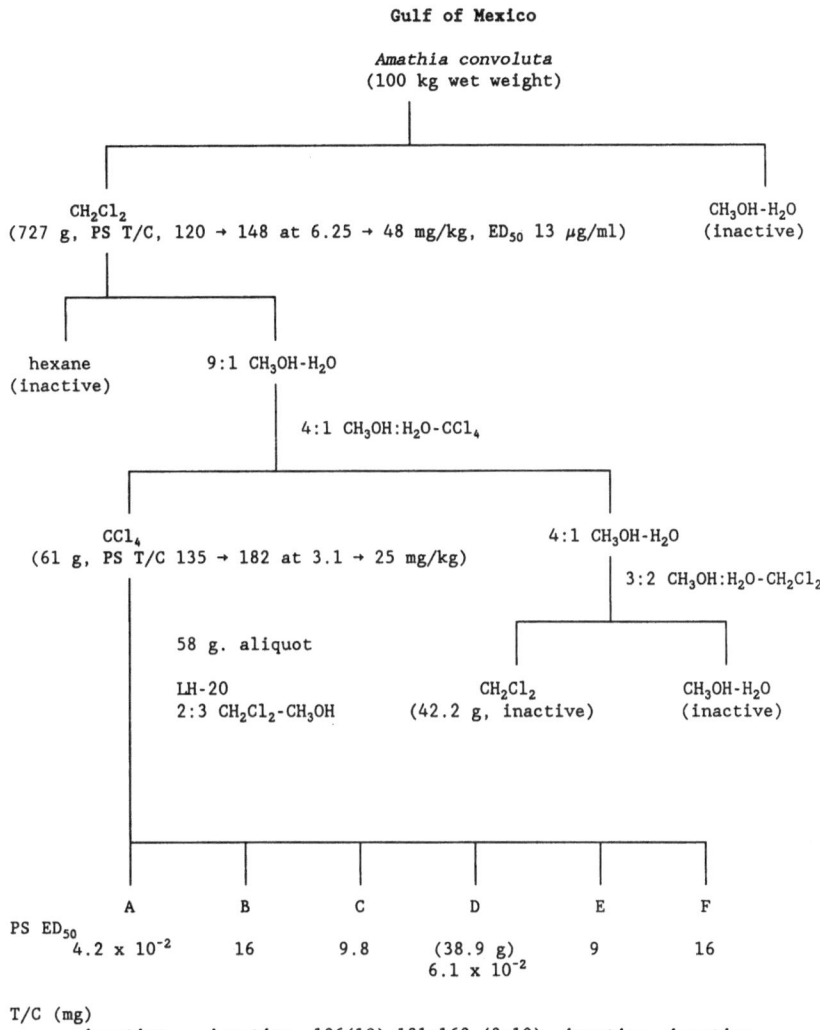

Scheme 1a. Fractionation of *Amathia convoluta* extract

produce a homogeneous solvent. The extraction was allowed to proceed and the overall procedure repeated as necessary. By this method the great mass of inactive materials is removed and the active fraction is reduced to a manageable weight.

The first two years spent on exploring the active fractions from *A. convoluta* were devoted to evaluating separation methods that ranged

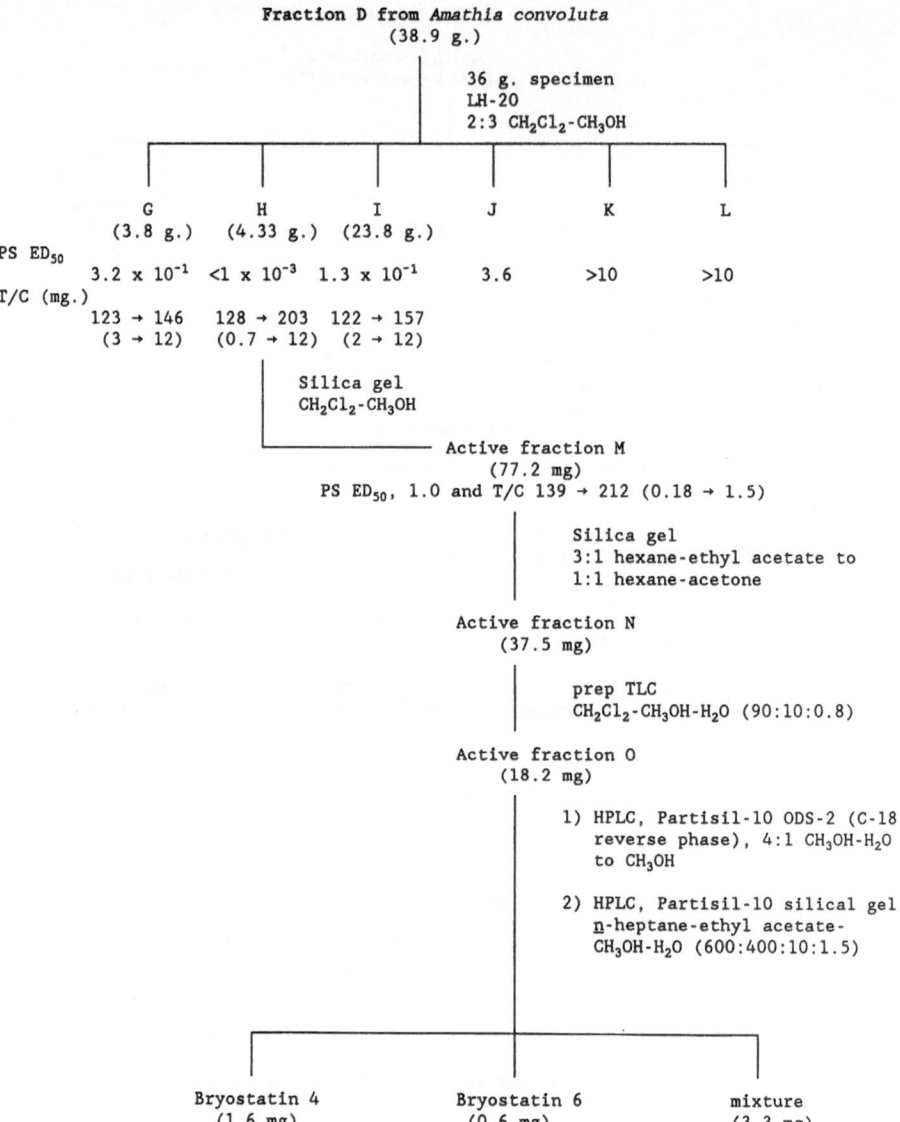

Fraction D from *Amathia convoluta*
(38.9 g.)

36 g. specimen
LH-20
2:3 CH_2Cl_2-CH_3OH

	G	H	I	J	K	L
	(3.8 g.)	(4.33 g.)	(23.8 g.)			
PS ED_{50}	3.2×10^{-1}	$<1 \times 10^{-3}$	1.3×10^{-1}	3.6	>10	>10
T/C (mg.)	$123 \to 146$	$128 \to 203$	$122 \to 157$			
	$(3 \to 12)$	$(0.7 \to 12)$	$(2 \to 12)$			

Silica gel
CH_2Cl_2-CH_3OH

Active fraction M
(77.2 mg)
PS ED_{50}, 1.0 and T/C $139 \to 212$ $(0.18 \to 1.5)$

Silica gel
3:1 hexane-ethyl acetate to
1:1 hexane-acetone

Active fraction N
(37.5 mg)

prep TLC
CH_2Cl_2-CH_3OH-H_2O (90:10:0.8)

Active fraction O
(18.2 mg)

1) HPLC, Partisil-10 ODS-2 (C-18
   reverse phase), 4:1 CH_3OH-H_2O
   to CH_3OH

2) HPLC, Partisil-10 silical gel
   n-heptane-ethyl acetate-
   CH_3OH-H_2O (600:400:10:1.5)

Bryostatin 4	Bryostatin 6	mixture
(1.6 mg)	(0.6 mg)	(3.3 mg)

Scheme 1b. Fractionation of *Amathia convoluta* extract

from gel permeation (Sephadex G-10 and LH-20), partition (Avicel A microcrystalline cellulose), ion exchange (Amberlite CG-120), and silica gel column chromatography to chromatography on macroreticular resins such as XAD-4. Of these methods use of Sephadex LH-20 at a

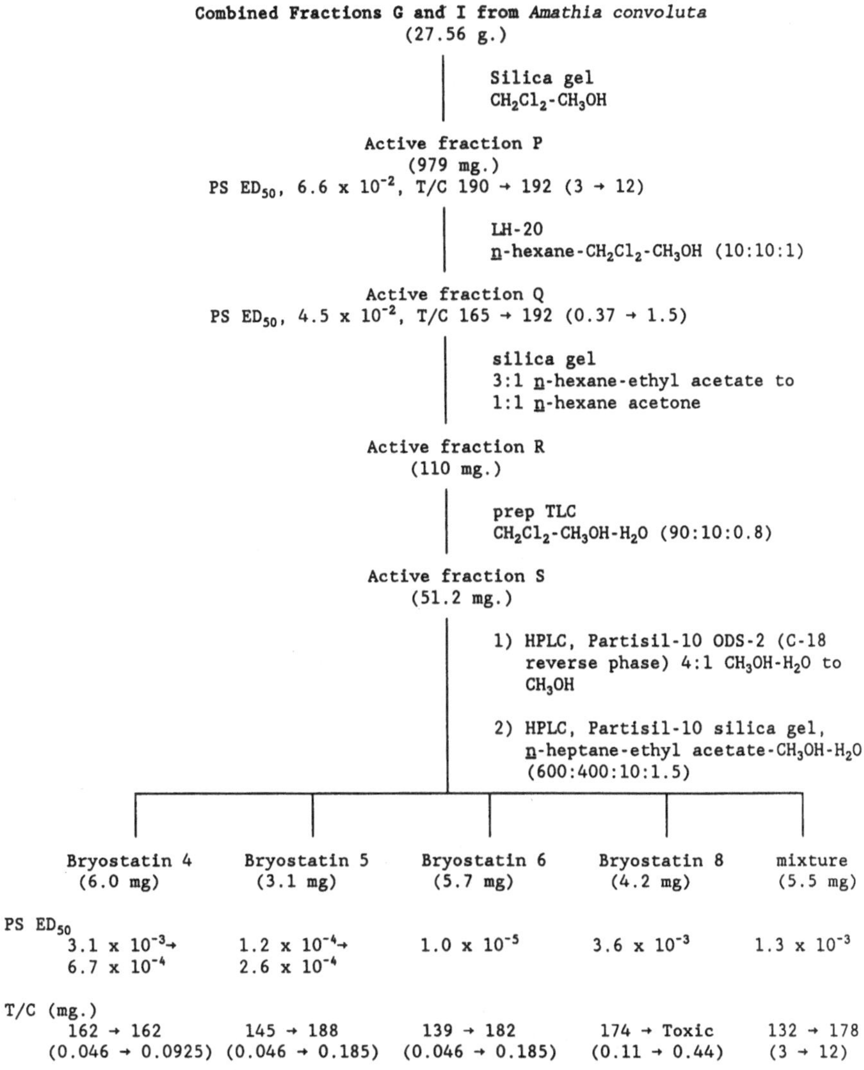

Combined Fractions G and I from *Amathia convoluta*
(27.56 g.)

| Silica gel
| CH_2Cl_2-CH_3OH

Active fraction P
(979 mg.)
PS ED_{50}, 6.6×10^{-2}, T/C $190 \to 192$ $(3 \to 12)$

| LH-20
| n-hexane-CH_2Cl_2-CH_3OH (10:10:1)

Active fraction Q
PS ED_{50}, 4.5×10^{-2}, T/C $165 \to 192$ $(0.37 \to 1.5)$

| silica gel
| 3:1 n-hexane-ethyl acetate to
| 1:1 n-hexane acetone

Active fraction R
(110 mg.)

| prep TLC
| CH_2Cl_2-CH_3OH-H_2O (90:10:0.8)

Active fraction S
(51.2 mg.)

1) HPLC, Partisil-10 ODS-2 (C-18 reverse phase) 4:1 CH_3OH-H_2O to CH_3OH

2) HPLC, Partisil-10 silica gel, n-heptane-ethyl acetate-CH_3OH-H_2O (600:400:10:1.5)

	Bryostatin 4 (6.0 mg)	Bryostatin 5 (3.1 mg)	Bryostatin 6 (5.7 mg)	Bryostatin 8 (4.2 mg)	mixture (5.5 mg)
PS ED_{50}	$3.1 \times 10^{-3} \to$ 6.7×10^{-4}	$1.2 \times 10^{-4} \to$ 2.6×10^{-4}	1.0×10^{-5}	3.6×10^{-3}	1.3×10^{-3}
T/C (mg.)	$162 \to 162$ $(0.046 \to 0.0925)$	$145 \to 188$ $(0.046 \to 0.185)$	$139 \to 182$ $(0.046 \to 0.185)$	$174 \to$ Toxic $(0.11 \to 0.44)$	$132 \to 178$ $(3 \to 12)$

Scheme 1c. Fractionation of *Amathia convoluta* extract

relatively early stage seemed most promising. As it became clear that the antineoplastic constituent(s) of *A. convoluta* must be present in only trace amounts it was decided in 1976 to recombine all fractions and to subject the combined material to the best separation technique developed to that time. But although the combined fractions still exhibited a high level of antineoplastic activity, it soon became apparent that the antineoplastic component(s) would continue to elude detection unless the quantities could be greatly increased. In 1978 approximately 118 kg (wet weight) was recollected followed by another 100 kg of the wet animal in early 1981. Experience gained from previous separation experiments was applied to the 1978 recollection and microgram amounts of the principal antineoplastic constituents were isolated over the next few years. The isolation procedures were further simplified, improved and applied to the 1981 recollection which forms the basis for the procedure outlined in Schemes 1a–c.

5. Chemical Transformations

A study of some chemical transformations was required to complete structure determinations of bryostatins 4–11. Use of 1% hydrochloric acid in methanol provided a method for fairly selective hydrolysis of the C-7 ester substituent. However, with bryostatin 10 (**10**) a new reaction pathway was uncovered. Key reactions and spectral interpretation used to deduce the structure of bryostatin 10 are illustrative (*28*) of our experience to that time.

Because of the seemingly close relationship of bryostatins 4 and 10, room temperature hydrolyses of the two compounds with 1% hydrochloric acid in methanol was expected to follow an analogous course. However, instead of the pivalate ester cleavage experienced by bryostatin 4, bryostatin 10 experienced only the loss of 1 mol of water. A new crystalline compound (0.32 mg) was obtained (from 1.2 mg of bryostatin 10) by high-performance liquid chromatography on a C-18 reversed-phase silica gel column using a gradient procedure with methanol–water mixtures. The SP-SIMS spectrum of the degradation product showed a molecular ion complex at m/z 790 (M^+) corresponding to loss of 18 mass units from bryostatin 10. Otherwise, an analogous SP-SIMS fragmentation was found. The ultraviolet spectrum exhibited maxima at 226 (ε 305 000) and 301 (ε 36 900) nm while the infrared spectrum of the dehydration product displayed new absorptions at 1660 and 1570 cm^{-1}. Most importantly, the ^{1}H-NMR signal at δ 2.441 originating in the H-20 protons of bryostatin 10 was absent. Thus, barring an undetected carbonium ion rearrangement the reaction appeared to have proceeded by elimination of the C-19 hydroxyl group and formation of a new C-19 → 20 double bond. The resulting diene system and extended chromophore from C-19 → C-35 seemed consistent with the ultraviolet spectrum and added further evidence for assignment of structure **10** to bryostatin 10. The driving force for dehydration probably resides in the delocalization energy resulting from this extended conjugated system. Interestingly, the dehydration product was also formed by allowing bryostatin 10 to remain in methylene chloride–methanol for 1 week at room temperature or for 30 min by adding a drop of 1% hydrochloric acid to a methylene chloride solution of starting material.

To provide further support for the structure assigned to bryostatin 10 a series of parallel acylation and oxidation experiments was performed with bryostatins 4 and 10. Mild acetylation (1 h at room temperature) of bryostatins 4 and 10 with acetic anhydride–pyridine gave monoacetate derivatives. In the ^{1}H-NMR spectra introduction of one acetyl group was apparent from appearance of a new sharp singlet at $\delta 2.056$. In both cases the H-26 signals usually seen at $\delta 3.77$ (m) and 3.74 (m) were shifted downfield to $\delta 4.99$ (m) and 5.036 (m), respectively. The ^{13}C-NMR spectra verified introduction of one acetate by new signals at $\delta 170.21$ and at $\delta 29.70$. Otherwise both proton and carbon spectra were as expected and almost identical with those of the parent substances. The SP-SIMS spectrum of bryostatin 10 26-acetate provided further verification. Such selectivity in the acetylation of bryostatins 4 and 10 at the C-26 hydroxyl group seems consistent with the x-ray crystal structure of bryostatin 1 which indicates that the C-3 hydroxyl group extends into the macrocyclic lactone cavity and is probably strongly hydrogen bonded, (see Fig. 4). Analogous selectivity was realized when bryostatins 4 and 10 were allowed to react with m-bromobenzoyl chloride in pyridine at ice-bath temperature. Again, interpretation of the NMR and SP-SIMS spectra established selective esterification of the C-26 hydroxyl group to yield the corresponding m-bromobenzoate esters. Selectivity was again observed when bryostatins 4 and 10 were oxidized with chromic acid–pyridine for one day at room temperature. The resulting ketones **13a** and **13b** (Fig. 3) gave analogous SP-SIMS spectra corresponding to formation of a monoketone and the ^{1}H-NMR spectra showed the C-27 methyl protons as sharp singlets at $\delta 2.153$ and 2.187, while the precursor C-26 proton was missing in the spectrum of each ketone.

Parallel results were found when bryostatins 4 and 10 were allowed to react with m-chloroperbenzoic acid in methylene chloride at room temperature. Introduction of one oxygen atom required two days. The SP-SIMS spectra were consistent with an additional 16 mass units, but while epoxidation of bryostatin 4 occurred at the Δ^{16}-olefin, epoxidation of bryostatin 10 occurred at the $\Delta^{13(30)}$-olefin, a conclusion reached by considering the ^{1}H-NMR spectra. The spectrum of bryostatin 10 epoxide retained the olefinic resonances of the Δ^{16} double bond but the H-30 olefinic resonance at $\delta 5.658$ was shifted to higher field at $\delta 3.358$, a shift characteristic of hydrogen on an epoxide. In addition, the C-36 methyl

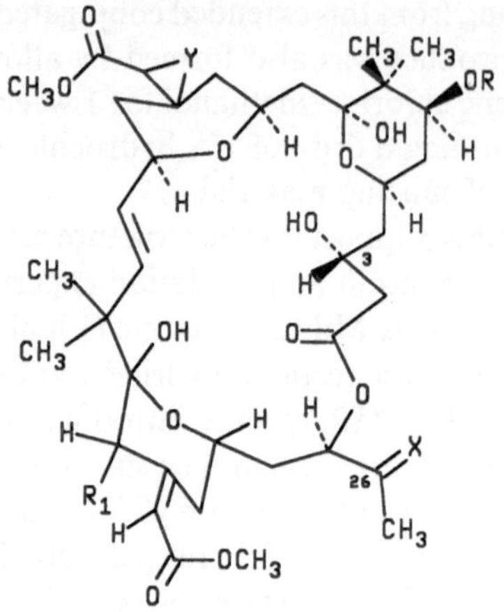

13a, R = COCH₂CH(CH₃)₄, R₁ = OCOCH₂CH₂CH₃, X = O, Y = π-bond

b, R = COCH₂CH(CH₃)₄, R₁ = H, X = O, Y = π-bond

c, R = COCH₂CH(CH₃)₄, R = OCOCH₂CH₂CH₃, X = H——, Y = O
                                                    HO

d, R = COCH₂CH(CH₃)₄, R₁ = H, X = H——, Y = O
                                        HO

Fig. 3. Transformation products of bryostatins 4 and 10

ester protons normally at δ3.687 were shifted slightly downfield to δ3.774. On the other hand, the product formed from bryostatin 10 exhibited the H-30 signal at δ3.359 compared with δ5.664 for H-30 of bryostatin 10 itself and the signal of the 36-methyl ester group was shifted to δ3.760. Unequivocal stereochemical assignments for the C-13 → C-30 epoxides (13c and 13d) and elimination of the possibility that epoxidation might have occurred at the C-21 → C-34 position will require an x-ray crystal structure determination.

To increase significantly the availability of bryostatin 1 for clinical trials it became very necessary to convert efficiently and selectively bryostatin 2 to bryostatin 1 since both compounds are obtained from *Bugula neritina* in nearly equal amounts. This objective was realized and also led to a practical synthetic method for preparing a series of new bryostatins. As noted above the bryostatin hydroxyl groups at C-3, C-9 and C-19 were found to resist acetylation under mild conditions, presumably due to intramolecular hydrogen bonding, while the C-7 and

C-26 hydroxyl groups were readily acetylated. Thus, conversion by methods utilizing selective protection of the C-26 hydroxyl group seemed possible. The steric environment of the two normally reactive hydroxyl groups (*viz*, C-7 and C-26) suggested use of a bulky silyl ether (*34*). Consequently, bryostatin 2 was allowed to react (*36*) at room temperature with excess *tert*-butyldimethylsilyl chloride (*35*, TBDMS) in the presence of 4-(*N*,*N*-dimethyl)aminopyridine (or triethylamine in dimethylformamide) to produce the corresponding 26-*tert*-butyldimethylsilyl (73.5% yield on the basis of total recovered bryostatin 2) and 7,26-di-*tert*-butyldimethylsilyl ethers. The disilyl ether was reconverted to bryostatin 2 employing (*37*) 48% hydrofluoric acid–acetonitrile (1:20). Treatment of the C-26 silyl ether with acetic anhydride–pyridine (room temperature) gave the C-7 acetate. The C-26 hydroxyl was regenerated using 48% hydrofluoric acid–acetonitrile (1:20 at 0–5°) to afford an 82% overall yield of bryostatin 1.

Selective protection of the C-26 hydroxyl group in bryostatin 2 allowed introduction of other groups at C-7. Treatment of 26-*tert*-butyldimethylsilyl ether with butyric anhydride and pyridine, followed by deprotection, led to bryostatin 2 7-butyrate identical with bryostatin 12 (**12**). Bryostatin 2 7-propionate was obtained in an analogous manner. Importantly, conversion of bryostatin 2 to bryostatin 2 7-isovalerate and bryostatin 2 7-pivalate provided convincing support for the revision (*38*) of the C-7 substituents in bryostatins 4 and 5, originally assumed to be isovalerates, to C-7 pivalates **4** and **5** respectively (*39*). The preceding strategy for selective reaction at the C-7 hydroxyl group of bryostatin 2 with protection and deprotection of the more reactive C-26 hydroxyl opened a useful pathway to a variety of new bryostatins for structure/activity studies. Other chemical transformations of these fascinating marine animal antineoplastic constituents are presently under study in our institute.

6. Absolute Configuration

In 1981 we were able to obtain x-ray quality crystals (space group $P2_12_12_1$) of bryostatin 1 from a layered solution of bryostatin 1 in methylene chloride under methanol (*17*). A total of 3553 useful reflections were obtained at $-100°C$ using graphite-monochromated M_oK_α (0.71069 Å) radiation and the crystal structure problem was solved employing MULTAN 78. The presumed correct enantiomer was selected by means of seven Bijvoet pairs with the largest ratios measured using CuK_α radiation (*17*). This approach led to the absolute configuration (**1**) proposed for bryostatin 1 in 1982, but we were unable to obtain a suitable heavy atom derivative analysis of which would confirm this assignment unequivocally until early 1990 (*40*).

Final elucidation of the absolute configuration of bryostatin 1 was achieved by converting (see Section 5) bryostatin 2 to a nicely crystalline C-7 *p*-bromobenzoate and obtaining 5279 (from a total of 13,707) usable reflections with CuK_α radiation. A combination of MULTAN 80 and SHELEX 5 86 methods led to the solution shown in Fig. 4. Among special features of this structure it should be noted that the three pyran rings are approximately in a chair conformation and each has a 4-position substituent that projects outward. All of the substituents are equatorial with respect to the pyran rings. An intramolecular hydrogen bond occurs between O-19H and O-3 (2.71, 2.71 Å), and two possible hydrogen bonds are found between O-3H and O-5 (2.84, 2.87 Å), and

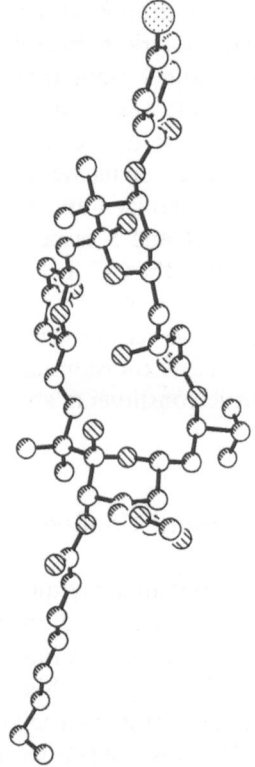

Fig. 4. ORTEP representation of bryostatin 2 7-*p*-bromobenzoate

between O-3H and O-11 (3.00, 3.02 Å). In the crystalline conformation oxygens O-1, O-3, O-5, O-11, O-19A, and O-19B are all on the interior of the large, oxygen-abundant cavity. The size and shape of this cavity and the arrangement of oxygen atoms suggested that the molecule might have cation-binding capabilities similar to those of the polyether antibiotics, but such properties have so far not been substantiated. The axial (E,E)-octa-2,4-dienoic acid substituent at C-20 would be expected to enhance lipid solubility and this substituent might swing over the internal cavity by rotation about the C-20-O-20 bond. The stereochemistry of the two acetylidene units at C-13 and C-21 is such that the carbonyl oxygen points in the direction of increasing carbon number along the macrocycle. While these units are disubstituted at the β-carbon, it is conceivable they might function as Michael acceptors for biopolymer amine and/or thiol groups.

Now that the absolute configuration of bryostatin 1 has been firmly established the stage is set for a better understanding of the biochemical role of the bryostatins (41) in cell biology and cancer chemotherapy.

7. Spectral and Analytical Characterization

The crystal structure determination (17) of bryostatin 1 in 1982 and the interpretation (29) of the corresponding high field (400 MHz) NMR spectra formed the basis for all subsequent bryostatin structure determinations. All 400 MHz ^{1}H-NMR spectra of the bryostatins until about 1986 were measured using an early model of the Bruker 400 instrument and required coupling assignments by hand. Hence some minor adjustments in the early proton assignments have already been found necessary.

As illustrations, Tables 2 and 3 list assignments for the ^{1}H-NMR spectra of two important bryostatins and two bryostatin acetates while Fig. 5 lists assignments for the ^{13}C-NMR spectrum of bryostatin 12. Fig. 6 shows the ultraviolet spectra, while mention of other ultraviolet, infrared, and mass spectral characteristics has already been made in Section 5. However an illustration of the application of mass spectrometry to the structure elucidation of bryostatins 5–7 will be instructive (21). The solution phase secondary ion mass spectra of these three bryostatins were quite revealing when new techniques for detecting molecular ions of hitherto refractory substances were utilized. By employing an alkali metal iodide such as sodium iodide or silver tetrafluoroborate or thallium tetrafluoroborate in sulfolane the molecular ions

Table 2. ¹H-NMR *Chemical Shift Assignments for Bryostatins 1 and 12* (400 MHz, CdCl₃)

H	Bryostatin 1 (c 4.4 mmol/L)			Bryostatin 12 (c 12.8 mmol/L)		
	δ	mult	$J_{H,H}$, Hz	δ	mult	$J_{H,H}$, Hz
2	2.46	brt	11	2.46	m	
2'	2.42	brd	11			
3	4.20	m		4.11	d	3
4a	2.02	m		2.01	m	
4b	1.58	d	8	1.56	d	8
5	4.23	m		4.23	d	12
6a	1.48	ddd	12.7, 12.7, 12.7	1.45	m	
6b	1.74	br dd	12.7, 5	1.72	m	
7	5.15	dd	12.7, 5	5.15	dd	12.2, 4.9
10a	2.07	br d	15.8	2.06	m	
10b	2.23	dd	7.5, 15.8	2.16	m	
11	3.83	m		3.85	m	
12a	2.07	m		2.07	dd	15.5, 7.2
12b	3.68	dd	16.5, 0.7	3.67	dd	15, 0.7
14a	1.88	m		1.88	m	
14b	3.67	dd	13.3, 1.5	3.66	dd	11.2, 1.5
15	4.07	dt	8.5, 2	4.07	dt	9.6, 2.2
16	5.30	dd	15.8, 8.5	5.30	dd	15.6, 8.3
17	5.79	d	15.8	5.78	d	15.6
20	5.18	s		5.18	s	
22a	2.08	br dd	15, 7.4	2.08	m	
22b	1.66	d	15	1.66	d	15
23	4.01	m		4.01	m	
24a	1.82	dd	11.5, 11.8	1.81	dd	11.3, 11.3
24b	1.97	dd	11.8, 11.8	1.95	dd	11.8, 11.8
25	5.17	ddd	14, 12.1, 3.0	5.17	ddd	14, 11.8, 2.9
26	3.78	m		3.78	m	
27	1.22	d	6.5	1.23	d	6.5
28[a]	0.94	s		0.94	s	
29[a]	1.00	s		1.00	s	
30	5.66	dd	1.3, 1.3	5.68	dd	2.2, 2.2
32[b]	1.15	s		1.15	s	
33[b]	1.00	s		1.00	s	
34	6.00	d	1.6	6.00	d	1.8
36	3.69	s		3.70	s	
37	3.66	s		3.66	s	
2'	2.05	s		2.28	t	7.5
3'				1.64	sextet	7.2
4'				0.94	t	7.3
2''	5.80	d	15.5	5.80	d	15.4
3''	7.25	m		7.25	m	
4''	6.16	br s		6.16	br s	
5''	6.16	br s		6.16	br s	
6''	2.15	dq	6.8, 3.0	2.14	dq	6.0, 2.5
7''	1.45	sextet	6.5	1.45	sextet	7.5
8''	0.92	t	7.6	0.92	t	7.3

Table 3. ^{1}H-NMR *Chemical Shift Assignments for Bryostatin 4 Acetate and Bryostatin 10 Acetate* (400 MHz, CdCl$_3$)

Assignment	Bryostatin 4 acetate		Bryostatin 10 acetate	
	δ	mult (*J*, Hz)	δ	mult (*J*, Hz)
2	2.43	m	2.46	m
3	4.13	m	4.14	m
4	1.60, 2.03	m, m	1.60, 2.05	m, m
5	4.20	m	4.17	m
6	1.40, 1.71	m, m	1.42, 1.74	m, m
7	5.09	m	5.06	m
10	1.50, 2.05	m	1.66, 2.16	m, m
11	3.80	m	3.84	m
12	2.20	m	2.18	m
14	1.88, 2.10	m, m	1.85, 2.05	m, m
15	4.10	m	4.12	m
16	5.28	dd (8.6, 15.8)	5.32	dd (8.4, 15.7)
17	5.76	d (15.8)	5.81	d (15.7)
20	5.16	s	2.43	d (10.3)
22	1.85, 2.00	m, m	1.85, 2.00	m, m
23	3.97	m	3.91	m
24	1.78, 1.90	m, m	1.78, 1.90	m, m
25	5.27	m	5.29	m
26	4.99	m	5.03	m
27	1.20	d (6.5)	1.24	d (5.9)
28[a]	1.12	s	1.05	s
29[a]	0.99	s	1.00	s
30	5.65	s	5.66	s
32[a]	0.97	s	1.00	s
33[a]	0.92	s	0.92	s
34	5.95	s	5.67	s
36	3.68	s	3.69	s
37	3.65	s	3.65	s
C-7 Pivalate				
2'	2.29	m	2.23	m
3'	~1.8–1.9	m	1.85–1.95	m
4' 5'	1.17	s	1.17	s
C-20 Butyrate				
2''	2.29	m		
3''	1.6–1.7	m		
4''	0.91	t (7.2)		
26-OCOCH$_3$	2.05	s	2.05	s

[a]Assignments for these four positions may be interchanged.

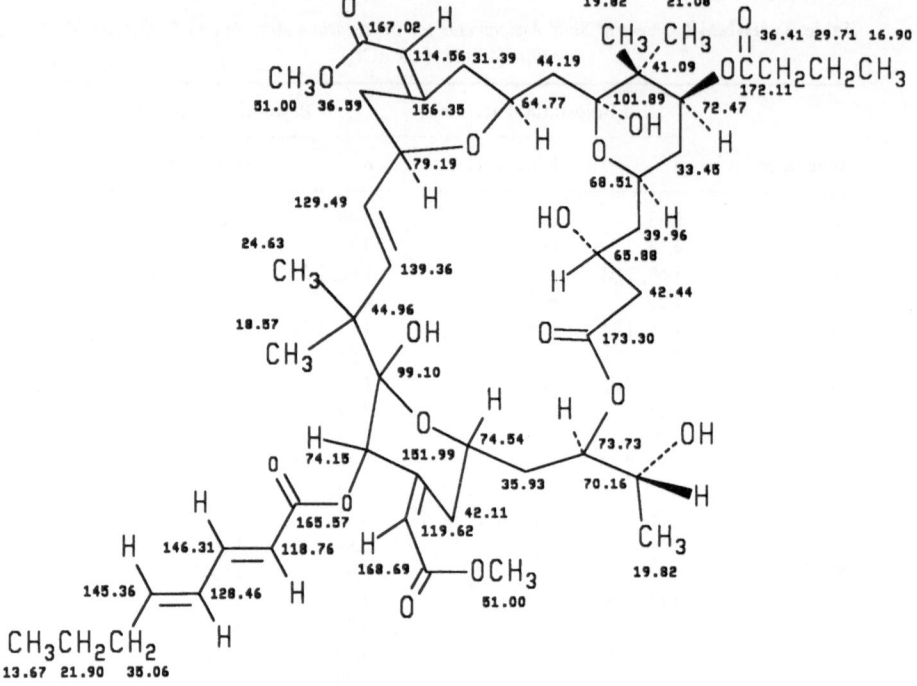

Fig. 5. ^{13}C-NMR assignments for bryostatin 12

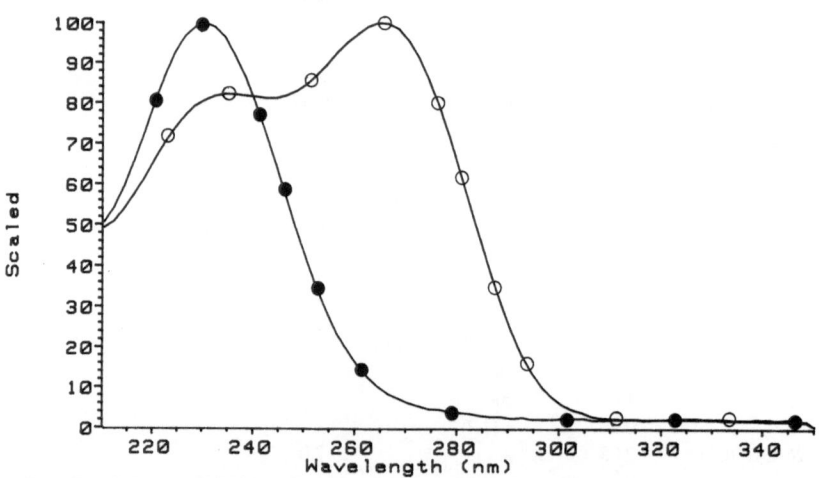

Fig. 6. UV spectra of bryostatins 1–3, 12 (———○———) and bryostatins 4–11 (———●———) recorded in acetonitrile–water (Photodiode array detector)

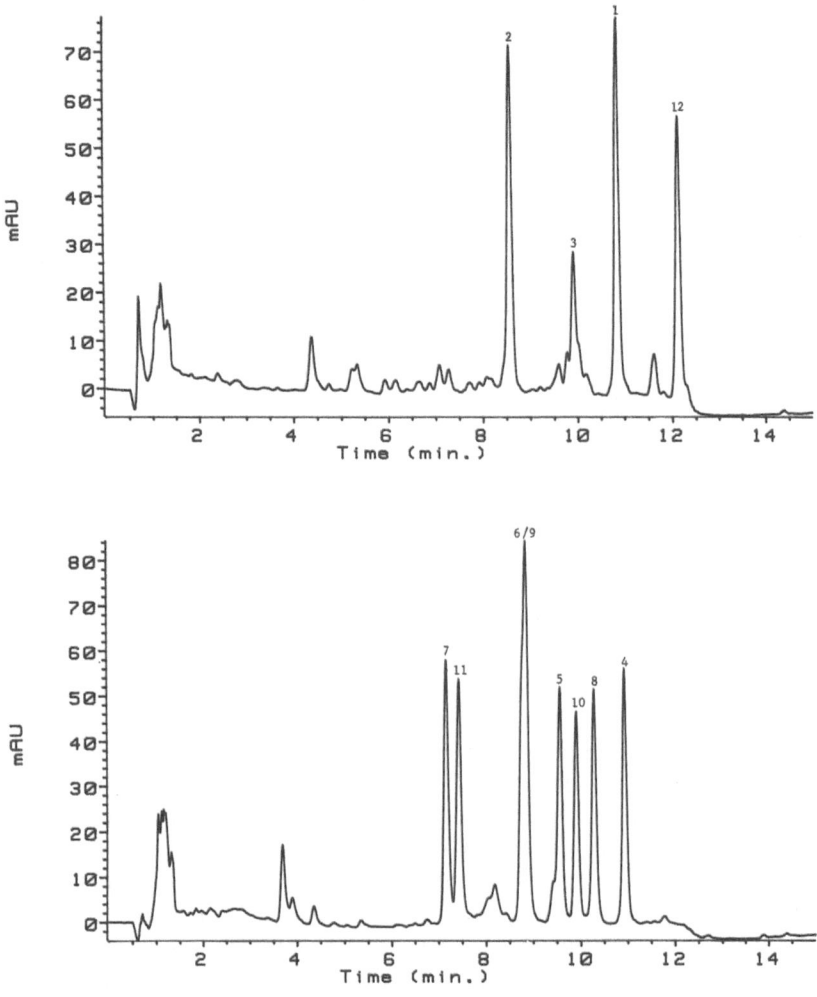

Fig. 7. HPLC-UV/vis (230 nm) of bryostatins 1–3, 12 (above) and 4–11 (below)

corresponding to bryostatins 5–7 were easily detected. Combination of this information with the few fragment ions produced proved to be very instructive; thus, the following results were obtained.

For bryostatin 5, 889 $[M + Na]^+$, 871 $[M + Na-18]^+$, 829 $[M + Na-60]^+$, and 788 $[M + Na-102]^+$; for bryostatin 6, 875 $[M + Na]^+$, 857 $[M + Na-18]^+$, 815 $[M + Na-60]^+$, and 787 $[M + Na-88]^+$; and for bryostatin 7, 847 $[M + Na]^+$, 829 $[M + Na-18]^+$, and 787 $[M + Na-60]^+$. Loss of 60 and 102 mass units by mass spectral fragmentation of bryostatin 5 suggested acetate and pivalate ester substituents. Similarly, loss of 60 and 88 mass units, from the bryostatin 6 molecular ion indicated acetate and butyrate esters. The cleavage of only 60 mass units from bryostatin 7 suggested a diacetate derivative.

For analytical characterization and assessment of purity a combination of high field (400 MHz) ^{1}H-NMR high resolution mass spectrometry (see above) and high performance liquid chromatography (HPLC) proved most reliable. For the latter a reversed-phase HPLC (RP-8; 1:1 to 100:0 acetonitrile–water gradient) separation procedure (38) was developed for detecting bryostatins 1–12, using a photodiode array detector system (cf. Fig. 7). While bryostatins 6 and 9 eluted together they were easily separated using a silica gel column with 9:1 n-hexane-1-propanol as eluent. By means of analogous HPLC procedures (42) the solubility (e.g., 0.77 mg/l in water and 97 mg/l in 60% ethanol–0.9% saline at 20°C) and stability (very stable at 0.1 mg/ml in ethanol at 20°C) of bryostatin 1 has been measured.

8. Antineoplastic and Cytostatic Activities

The biological activity of bryostatin 1 (1) was found to be exceptional. Against the murine PS leukemia this macrocyclic lactone showed 52–96% life extension at 10–70 µg/kg (injection/dose) levels, and an ED_{50} of 0.89 µg/ml against the P388 in vitro cell line. With the murine L-1210 lymphocytic leukemia 34–51% life extension was found at 37.5–150 µg/kg. Against the NCI murine M5 ovarian carcinoma tumor regression model a 20–65% curative response was seen at 20–40 µg/kg. Next, bryostatin 1 was found to afford 31–68% life extension at 5–40 µg/kg against the M531 murine ovary sarcoma (ip tumor implant with ip treatment). Most importantly, bryostatin 1 has recently shown remarkable selectivity (13) against human cancer cell lines (NCI) representing leukemias, renal cancer (A704), melanoma (SK-MEL-5), and non-small cell lung cancer (A549). Bryostatin 2 (2) has proved to be comparably active showing, e.g., a 60% increase in life-span at 30 µg/kg against the PS leukemia. Bryostatin 3 (3) showed a 63% life extension at 30 µg/kg in the PS leukemia.

More recently it has been found (unpublished experiments) in the laboratories of Dr. Peter BLUMBERG (NCI) that bryostatins 1 and 2 will afford PS ED_{50} values in the 10^{-4} µg/ml range if the leukemia cells are exposed to only that concentration range. In other words the cytostatic activity of the bryostatins markedly increases as their concentration decreases. Furthermore Dr. Michael BOYD (NCI) has discovered that the bryostatins elicited a remarkable increase in activity against a variety of human cell lines (*13*) when evaluation was delayed from two to six days. At Johns Hopkins University Dr. Stratford MAY has reexamined the original NCI *in vivo* murine B-16 melanoma experiments and found bryostatin 1 capable of curing this tumor. Since bryostatin 3 (**3**) retains the very potent antineoplastic activity of bryostatin 1 (**1**), loss of one pyran acetylidene group does not appear critical. Indeed, present evidence suggests that substantial structural modifications can be made in the bryostatin ester substituents while preserving antineoplastic activity. The discovery of bryostatin 4 (**4**) with pronounced cell growth (PS cell line ED_{50}, 10^{-3}–10^{-4} µg/ml) and antineoplastic (PS, 62% increase in life extension at 46 µg/kg) properties eliminated the (*E,E*)-octa-2,4-dienate at C-20 of bryostatins 1–3 as essential for such important biological properties.

The antineoplastic activity of bryostatins 5–7 was again impressive as follows: bryostatin 5, PS cell line ED_{50}, 1.3×10^{-3} to 2.6×10^{-4}, PS *in vivo* 88% life extension at 185 µg/kg; bryostatin 6, PS cell line ED_{50}, 1.0×10^{-5} µg/ml, *in vivo*, 82% life extension at 185 µg/kg, and bryostatin 7, PS cell line ED_{50}, 2.6×10^{-5} µg/ml, *in vivo* 77% life extension at 92 µg/kg. Bryostatin 8 gave 74% life extension at 110 µg/kg against the PS leukemia. Bryostatin 9 (NSC 606639), as usual, significantly inhibited the PS leukemia showing 40% life extension at 80 µg/kg and cell line ED_{50} 1.2×10^{-3} µg/kg. Interestingly, the reversal of ester groups in bryostatins 6 and 9 did not appear to affect strongly the PS activity. Both bryostatins 10 (**10**) and 11 (**11**) proved to be markedly active against the PS leukemia exhibiting PS cell growth inhibition at ED_{50} 2.6×10^{-4} and 1.8×10^{-5} µg/ml and PS *in vivo* growth inhibition at, for example, 34% at 10 µg and 64% at 92.5 µg/kg respectively. Bryostatin 12 (NSC 606294) exhibited activity against the PS leukemia cell line at ED_{50} 0.014 µg/ml and gave a 47–68% life extension at 30–50 µg/kg doses, while bryostatin 13 (NSC 606740) led to ED_{50} 0.0054 µg/ml. The antineoplastic properties of bryostatins 10 and 11 eliminate an oxygen substituent at C-20 at a prerequisite for such important biological properties. The presence of both bryostatins 10 and 11 in specimens of *B. neritina* from the Gulfs of Mexico and California indicates that these new bryostatins will be found in other geographically diverse specimens of this intriguing bryozoan. Indeed present evidence suggests that additional (and important) structure/activity information will be obtained by isolating other new antineoplastic substances produced by the versatile *B. neritina*, by chemical transformations of the bryopyran substituents (and ring system) and by evaluating the intermediates employed in approaches to total synthesis.

In the latter regard a considerable number of expert synthetic groups have been attracted to the challenges offered by the bryostatins. At the time of this writing (February, 1990) none had completed the total synthesis of a bryostatin, but two were very close. The group of S. MOSAMUNE has announced (43) a synthesis of the seco-acid corresponding to bryostatin 7 and the group of DAVID EVANS (44) was close to coupling four segments needed to complete a bryostatin.

9. Inhibition of Tumor Promotion

At present, the enzyme system discovered in 1977 known as protein kinase C (45) is believed to comprise a large series of related proteins with subtle and specialized enzymological properties. The enzyme system is activated by a diacylglycerol that arises from a receptor controlled hydrolysis of inositol phospholipids. The protein kinase C (PKC) in turn regulates many intracellular processes by relaying information from extracellular signals across the cell membrane. Each member of the PKC enzyme series has a special role in regulating membrane functions and in the activation of gene transcription. In turn this results in a great variety of very important cellular responses such as release of insulin, growth hormone, luteinizing hormone, prolactin, thyrotropin, parathyroid hormone, calcitonin, dopamine, serotonin, histamine and superoxide as well as secretion of aldosterone, amylase, catecholamine, pepsinogen and T- and B-lymphocyte activation. PKC is most abundant in brain cells and is very important in cell-to-cell communication, growth control, cell differentiation, and, e.g., memory formation. Most pertinent to this review is the fact that PKC has been strongly implicated in tumor promotion and carcinogenesis (46).

In the 1984 period while attempting to ascertain whether or not bryostatin 1 was an ionophore we discovered that it might have a potentially important effect on the PKC system (47) when it inhibited the binding of ^{3}H-phorbol to a cellular high affinity receptor. This observation raised the unsettling possibility that bryostatin 1 like some of the phorbol plant constituents might be a tumor promoter. Thanks to the very expert investigations and intense interest of Dr. PETER BLUMBERG (NCI) we found instead that bryostatin 1 is a potent *anti*tumor promoter and will even inhibit (46) the tumor promoting (47) activities of the phorbols in mouse keratinocytes (48) and Sencar mouse skin (49, 50). In a parallel investigation analogous results were obtained by GSCHWENDT and colleagues (51) in the German Cancer Research Center in Heidelberg. These studies stimulated a number of related investigations concerned with intermediate biochemical pathways (*52–60*) and the relationship of bryostatins 1 and 4 to the phorbol ester pharmacophore (*61*). The latter study using molecular modeling based on the bryostatin x-ray coordinates suggested that the C-20, C-9, and C-4 oxygens of phorbol and the C-1, C-19, and C-26 oxygen atoms of the bryostatins were elements critical to the pharmacophore.

10. General Biological Properties

10.1. Immunological

By 1987 we had substantial evidence that bryostatins 1 and 2 reacted in a synergistic fashion with recombinant B-cell stimulatory factor/interleukin 4 to stimulate resting T-cells to proliferate and differentiate into cytotoxic T-lymphocytes (*62*). At the same time interleukin-2 (IL-2) development of cytotoxicity was greatly enhanced (*62, 63*) suggesting that the bryostatins may allow the clinical dose of the hazardous recombinant IL-2 to be considerably reduced. In a related study concerned with the effects of bryostatin 1 on IL-2 and γ-interferon synthesis by ionophore A23187 or mitogen-induced human blood lymphocytes we found the production of these two lymphokines to be increased 10–100 fold (*63*). Interestingly, bryostatin 1 was also found (*64*) to effectively block T-cell proliferation induced by a phorbol (PMA).

10.2. Hematopoietic

Mammalian blood is an amazingly versatile and necessary biosynthetic product of evolution. All blood cells are derived from a small number of pluripotent stem cells that serve as a single progenitor. For example, the erythrocytes and the leukocytes are all formed and regulated by a complex set of cytokines that help direct hematopoietic (from the Gr. *haima* for blood and *poiein*, to make) progenitors at different stages of development (*65, 66*). The immune system leukocytes include the specialized lines of granulocytes (which in turn include the sub-sets of basophils, eosinophils that attack protozoans and worms, and neutrophils which defend against bacteria and some fungi), lymphocytes (which destroy a wide range of pathogens) and monocytes (and related macrophages which defend against some bacteria, viruses and other intracellular parasites).

Presumably the original stem cell synthesizes surface receptors that detect certain hormone signals directing the cell onward to its final specialized task. The first important branching affords the lymphocyte precursors and as differentiation proceeds the erythroblasts (which produce red cells) and the myeloblasts (which produce granulocytes and monocytes) are formed. Hematopoietic growth factors (HGF) comprise the colony-stimulating factors (CSF) and interleukins (IL). Overall this is a group of glycoproteins with a mass range of 18,000–90,000 (67) that regulate differentiation and proliferation of the various cell lineages. The first HGF, erythropoietin which is used to treat anemia associated with end-stage kidney disease by avoiding repeated blood transfusions, stimulates proliferation of erythroid progenitors, and induces hemoglobin formation, was discovered in 1906 (67), but it took some 60 years until other HGFs began to be uncovered. Now twelve such proteins are known and available, albeit in questionable purity, by cloning techniques. These include (66) IL-2 (a co-factor for growth and differentiation of B- and T-cells which augments lymphocyte-activated killer activity and induces production of other lymphokines), IL-3 (which promotes early growth of granulocyte, monocyte, erythroid and megakaryocyte progenitor cells as well as mast cell growth and induces leukemia blasts to proliferate) and granulocyte-macrophage colony-stimulating factor (GM-CSF) whose action is similar to that of IL-3 and which like IL-3 can be used clinically for bone marrow transplants, also for drug-induced neutropenia, leukopenia in AIDS and aplastic anemia. Doubtlessly, the colony-stimulating factors will find increasing clinical applications if their production of potent side-products such as with IL-1, interferons, fibroblast growth factor and tumor necrosis factor (cockectin) can be moderated (67). Presently the bryostatins seem to be more easily controlled and a clinically useful alternative for certain IL-2, IL-3, GM-CSF and other HGF applications.

In early 1987 we proved that the bryostatins were capable of stimulating normal bone marrow progenitor cells to form colonies *in vitro* and to activate neutrophils (68). Bryostatin 1, e.g., promotes many of the biological effects of GM-CSF and this remarkable activity combined with its antineoplastic activity make it a very attractive clinical candidate. Meanwhile, bryostatins 1 and 2 have found an important role as biochemical probes for unraveling the mechanisms of normal hematopoiesis (69). An important advance here was the observation that bryostatin 1 will stimulate normal erythropoiesis in human bone marrow progenitor assays. Furthermore, bryostatin 1 approximated the stimulatory effects of IL-3 on both murine normal and w/w bone marrow (deficient in T-lymphoid cells) cells. Release of IL-3 by bryostatin 1 was believed responsible for the stimulation of erythropoiesis (*69*).

10.3. Microbiological

Preliminary evaluation of bryostatins 1 and 2 against several microbial systems proved to be negative. However, bryostatin 1 was found to inhibit induction of Epstein-Barr virus in Raji cells, a potentially important lead along with its GM-CSF and IL-3 activity to a possible application in HIV-1 therapy. However, only a clinical trial will provide a definitive answer to this proposal.

11. Mechanism of Action

As circumstances dictate the bryostatins appear to either activate or in some cases inhibit members of the PKC enzyme family. Prior to the excellent studies of BLUMBERG and co-workers (*46*) activation of PKC typical of phorbol esters was considered a requirement for tumor promotion. Bryostatins 1 and 2 were shown to be exceptions to this concept. In general PKC seems to have an enzymatic site responsible for kinase activity and a regulatory area that binds calcium, phospholipid, and, e.g., phorbol esters. The antagonism to phorbol ester action by bryostatins was found (*46*) to be non-competitive and suggested a different high-affinity site for the bryostatins and/or reaction with a different sub-set of the PKC isoenzymes. Interestingly, the dissociation constant for bryostatin 1 with PKC was located in the picomolar range rather than the nanomolar concentrations typical of the phorbols. But by increasing the concentration one thousand fold certain phorbol esters can induce PKC to phosphorylate the same HL-60 cellular proteins as experienced using bryostatin 1 (*46*). The next clue to the mechanism of action of the bryostatins was obtained (*70*) when bryostatin 1 induced phosphorylation of the major nuclear envelope polypeptide lamin B involved in depolymerization during mitosis.

Obviously, the PKC isoenzymes and the general complexity of this elaborate enzyme system severely complicate the bryostatin mechanistic problem (*71–74*), at this time. The facts are further obscured by results from different laboratories employing a well-known cell line such as HL-60 which proved to have previously unrecognized sub-lines (*73, 74*).

12. Implications for Cancer Treatment

Research results summarized in Sections 8–10 strongly indicate that bryostatin 1 and/or others in the series may prove to be clinically useful for treatment of cancer and certain other medical problems. Recently, bryostatin 1 has been found to inhibit six of seven fresh human myeloid leukemia specimens (74), prevent (whereas the phorbol PMA induces) squamous differentiation in human tracheobronchial epithelial cells (75), inhibit A549 human lung carcinoma cells (76) and very importantly induce differentiation of human B-chronic lymphocytic leukemia (CLL) cells (77, 78). The latter important contribution by DREXLER and colleagues may open a clinical path to successfully treating the routinely fatal CLL.

13. Conclusions

The systematic discovery and development of new drugs for cancer chemotherapy and other medical problems based on special metabolites from animal and plants has been seriously impeded by a general lack of understanding of its great potential, a situation which has been greatly aggravated by the paucity of organic chemists being trained for such very challenging research and the continuing lack of necessary financial support for research personnel and laboratories. The story of the bryostatins provides a sparkling glimpse of the potential for medical, biological and chemical discoveries based on constituents of marine organisms. In addition to the purely chemical challenge provided by the bryostatins their antineoplastic, *anti*tumor promoting, erythropoiesis regulating, hematopoietic stimulating (GM-CSF) and immunomodulating properties give promise of improving cancer treatment and perhaps other medical problems. It is also interesting that except for some fish, the California *Bugula neritina* has only one known predator, *Polycera atra*. If the bryostatins are in turn utilized for defensive and/or regulatory purposes by this animal, its common zoological description as the sorcerer's nudibranch will prove to be brilliantly prescient.

Acknowledgement

The ASU-CRI research on the bryostatins was supported by the NCI Outstanding Investigator Grant OIG CA44344-01A1, Grant CA-16049-10-12 awarded by the Division of Cancer Treatment, NCI, DHHS, the National Cooperative Drug Discovery Grant No.

References, pp. 191–195

AI-25696-02, the Fannie E. Rippel Foundation, the Arizona Disease Control Research Commission, the Robert B. Dalton Endowment Fund, Eleanor W. Libby, the Waddell Foundation (Donald Ware) and Herbert K. and Dianne Cummings (The Nathan Cummings Foundation, Inc.). We also appreciate the expert assistance of Drs. Feng Gao, Cherry L. Herald, Fiona Hogan and Yoshiaki Kamano and Mrs. Christine Duplissa.

References

1. KORNBERG, A.: The Two Cultures: Chemistry and Biology. Biochemistry 26, 6888 (1987).
2. HARSHBARGER, J.C.: Invertebrate animals—what can they contribute to cancer research? Federat. Proc. (Amer. Soc. Exp. Biol.) 32, 2224 (1973).
3. HARSHBARGER, J.C., and C.J. DAWE: Hematopoietic Neoplasms in Invertebrate and Poikilothermic Vertebrate Animals. In: Unifying Concepts of Leukemia, Bibl. haemat. No. 39, (DUTCHER, R.M., and L. CHIECO-BIANCHI, eds.), p. 1. Basel: Karger. 1973.
4. PETTIT, G.R., J.F. DAY, J.L. HARTWELL, and H.B. WOOD: Antineoplastic Components of Marine Animals. Nature 227, 962 (1970).
5. PETTIT, G.R., J.L. HARTWELL, and H.B. WOOD: Arthropod Antineoplastic Agents. Cancer Res. 28, 2168 (1968).
6. PETTIT, G.R.: Biosynthetic Products for Cancer Chemotherapy, Vol. 1, p. 165. New York: Plenum Press. 1977.
7. PETTIT, G.R., Y. KAMANO, P. DRASAR, M. INOUE, and J.C. KNIGHT: Synthesis of Bufalitoxin and Bufotoxin. J. Organ. Chem. (USA) 52, 3573 (1987).
8. PETTIT, G.R., L.E. HOUGHTON, N.H. ROGERS, R.M. COOMES, D.F. BERGER, P.R. REUCROFT, J.F. DAY, J.L. HARTWELL, and H.B. WOOD, Jr.: Butterfly Wing Antineoplastic Agents. Experientia 28, 382 (1972).
9. PETTIT, G.R., and R.H. ODE: Antineoplastic Agents. 41. The Beetle *Allomyrina dichotomus*. Lloydia 39, 129 (1976).
10. PETTIT, G.R., R.M. BLAZER, and D.A. REIERSON: Antineoplastic Agents. 51. The Yellow Jacket *Vespula pensylvanica*. Lloydia 40, 247 (1977).
11. PETTIT, G.R., R.H. ODE, and T.B. HARVEY, III: Isolation of Taurine from the Molluscs *Macrocallista nimbosa* and *Turbo stenogyrus*. Lloydia 36, 204 (1973).
12. PETTIT, G.R., C.L. HERALD, and D.L. HERALD: Antineoplastic Agents XLV: Sea Cucumber Cytotoxic Saponins. J. Pharm. Sci. 65, 1558 (1976).
13. SUFFNESS, M., D.J. NEWMAN, and K. SNADER: Discovery and Development of Antineoplastic Agents from Natural Sources. In: Bioorganic Marine Chemistry, Volume 3, p. 131. Berlin-Heidelberg-New York-Tokyo Springer. 1989.
14. Biology of Bryozoans (Woollacott, R.M., and R.L. ZIMMER, eds.). New York: Academic Press. 1977.
15. MCKINNEY, F.K., and J.B.C. JACKSON: Bryozoan Evolution. In: Special Topics in Palaeontology, Volume 2, xiv, 238 pp. Winchester, MA: Unwin Hyman. 1989.
16. RYLAND, J.S., and P.J. HAYWARD. British Anascan Bryozoans, Cheilostomata: Anasca. In: Synopses of the British Fauna No. 10, (KERMACK, D.M., ed.), p. 1. London: Academic Press. 1977.
17. PETTIT, G.R., C.L. HERALD, D.L. DOUBEK, D.L. HERALD, E. ARNOLD, and J. CLARDY: Isolation and Structure of Bryostatin 1. J. Amer. Chem. Soc. 104, 6846 (1982).
18. PETTIT, G.R., C.L. HERALD, Y. KAMANO, D. GUST, and R. AOYAGI: The Structure of Bryostatin 2 from the Marine Bryozoan *Bugula neritina*. J. Nat. Prod. 46, 528 (1983).

19. PETTIT, G.R., C.L. HERALD, and Y. KAMANO: Structure of the *Bugula neritina* (Marine Bryozoa) Antineoplastic Component Bryostatin 3. J. Organ. Chem. USA **48**, 5354 (1983). See also ref. 40 below.

20. PETTIT, G.R., Y. KAMANO, C.L. HERALD, and M. TOZAWA: Structure of Bryostatin 4. An Important Antineoplastic Constituent of Geographically Diverse *Bugula neritina* (Bryozoa). J. Amer. Chem. Soc. **106**, 6768 (1984).

21. *Idem.* Isolation and Structure of Bryostatins 5–7. Canad. J. Chem. **63**, 1204 (1985).

22. PETTIT, G.R., Y. KAMANO, R. AOYAGI, C.L. HERALD, D.L. DOUBEK, J.M. SCHMIDT, and J.J. RUDLOE: Antineoplastic Agents 100. The Marine Bryozoan *Amathia convoluta*. Tetrahedron **41**, 985 (1985).

23. PETTIT, G.R., J.E. LEET, C.L. HERALD, Y. KAMANO, and D.L. DOUBEK: Antineoplastic Agents, 116. An Evaluation of the Marine Ascidian *Aplidium californicum*. J. Nat. Prod. **49**, 231 (1986).

24. MORRIS, R.H., D.P. ABBOTT, and E.C. HADERLIE, Intertidal Invertebrates of California, p. 177 and plate 57. Stanford, CA: 1980.

25. PETTIT, G.R., Y. KAMANO, and C.L. HERALD: Antineoplastic Agents, 118. Isolation and Structure of Bryostatin 9. J. Nat. Prod. **49**, 661 (1986).

26. PETTIT, G.R., C.W. HOLZAPFEL, and G.M. CRAGG: Mass Measurements of Natural Products by Solution Phase Secondary Ion Mass Spectrometry Employing Silver(I) and Thallium(I) Derivatives. J. Nat. Prod. **47**, 941 (1984).

27. PETTIT, G.R., C.W. HOLZAPFEL, G.M. CRAGG, C.L. HERALD, and P. WILLIAMS: Broad Scope Secondary Ion Mass Spectrometry. J. Nat. Prod. **46**, 917 (1983).

28. PETTIT, G.R., Y. KAMANO, and C.L. HERALD: Isolation and Structure of Bryostatins 10 and 11. J. Organ. Chem. USA **52**, 2848 (1987).

29. ARENE, E.O., G.R. PETTIT, and R.H. ODE: The Isolation of Isosakuranetin Methyl Ether from *Eupatorium odoratum*. Lloydia **41**, 186 (1978).

30. BLIGH, E.G., and W.J. DYER: A Rapid Method of Total Lipid Extraction and Purification. Canad. J. Biochem. Physiol. **37**, 911 (1959).

31. PETTIT, G.R., Y. FUJII, J.A. HASLER, J.M. SCHMIDT, and C. MICHEL: Antineoplastic Agents 78. Isolation of Palystatins 1–3 from the Indian Ocean *Palythoa liscia*. J. Nat. Prod. **45**, 263 (1982).

32. PETTIT, G.R., J.E. LEET, C.L. HERALD, Y. KAMANO, F.E. BOETTNER, L. BACZYNSKYJ, and R.A. NIEMAN: Isolation and Structure of Bryostatins 12 and 13. J. Organ. Chem. (USA) **52**, 2854 (1987).

33. PETTIT, G.R., Y. KAMANO, C.L. HERALD, J.M. SCHMIDT, and C.G. ZUBROD: Relationship of *Bugula neritina* (Bryozoa) antineoplastic constituents to the yellow sponge *Lissodendoryx isodictyalis*. Pure & Appl. Chem. **58**, 415 (1986).

34. KOCIENSKI, P.J., D.A.S. STREET, C. YEATES, and S.F. CAMBELL: The 3,4-Dihydro-2H-pyran Approach to (+)-Milbemycin β₃. Part 2. An Improved Synthesis of (2R,4S,6S,8R,9S)-2-[(5R)-(2E)-3-Methyl-5-formyl-hex-2-en-1-yl]-8,9-dimethyl-4-(dimethyl-*t*-butylsilyloxy)-1,7-dioxaspiro[5.5]-undecane. J. Chem. Soc. Perkin Trans I **1987**, 2189.

35. COREY, E.J., and A. VENKATESWARLU: Protection of Hydroxyl Groups as *tert*-Butyldimethylsilyl Derivatives. J. Amer. Chem. Soc. **94**, 6190 (1972).

36. LALONDE, M., and T.H. CHAN: Use of Organosilicon Reagents as Protective Groups in Organic Synthesis. Synthesis **1985**, 817.

37. ZIEGLER, F.E., A. NANGIA, and G. SCHULTE: The Synthesis of Neosporol: A Trichothecene in Search of a Natural Product. Tetrahedron Letters **29**, 1669 (1988).

38. PETTIT, G.R., Y. KAMANO, D. SCHAUFELBERGER, C.L. HERALD, P.M. BLUMBERG, and W.S. MAY: High Performance Liquid Chromatographic Separation of Bryostatins 1–12. J. Liquid Chromatog. **12**, 553 (1989).

39. PETTIT, G.R., D. SENGUPTA, C.L. HERALD, and P.M. BLUMBERG: Synthetic Conversion of Bryostatin 2 to Bryostatin 1 and Related Bryopyrans. *Can. J. Chem.* in press.
40. PETTIT, G.R., D.L. HERALD, F. GAO, D. SENGUPTA and C.L. HERALD: Absolute Configuration of the Bryostatins. J. Organ. Chem. (USA) **56**, 1337 (1991).
41. GIGNAC, S.M., M. BUSCHLE, G.R. PETTIT, A.V. HOFFBRAND, and H.G. DREXLER: Differential Expression of Trap Isoenzyme in B-CLL Cells Treated with Different Inducers. Leukemia & Lymphoma **3**, 19 (1990).
42. BAER, J.C., J.A. SLACK, and G.R. PETTIT: Stability-indicating high-performance liquid chromatography assay for the anticancer drug bryostatin 1. J. Chromatog. **467**, 332 (1989).
43. KAGEYAMA, M., and S. MASAMUNE: Studies Toward the Total Synthesis of Bryostatin 7. 31st Symposium on the Chemistry of Natural Products, Nagoya, Japan, **1989**, 227.
44. EVANS, D.A., E.M. CARREIRA, A.B. CHARETTE, and J.A. GAUCHET: Approaches to the Synthesis of Antineoplastic Macrolide Bryostatin. Amer. Chem. Soc. National Meeting, Dallas, Texas, April 10, 1989.
45. Y. NISHIZUKA: The Family of Protein Kinase C for Signal Transduction. J. Amer. Med. Assoc. **262**, 1826 (1989).
46. BLUMBERG, P.M., G.R. PETTIT, B.S. WARREN, A. SZALLASI, L.D. SCHUMAN, N.A. SHARKEY, H. NAKAKUMA, M.L. DELL'AQUILA, and D.J. DE VRIES: The Protein Kinase C Pathway in Tumor Promotion. In: Skin Carcinogenesis: Mechanisms and Human Relevance, p. 201. Alan R. Liss, Inc. 1989.
47. SMITH, J.B., L. SMITH, and G.R. PETTIT: Bryostatins: Potent, New Mitogens that Mimic Phorbol Ester Tumor Promoters. Biochem. Biophys. Res. Comm. **132**, 939 (1985).
48. SAKO, T., S.H. YUSPA, C.L. HERALD, G.R. PETTIT, and P.M. BLUMBERG: Partial Parallelism and Partial Blockade by Bryostatin 1 of Effects of Phorbol Ester Tumor Promoters on Primary Mouse Epidermal Cells. Cancer Res. **47**, 5445 (1987).
49. HENNINGS, H., P.M. BLUMBERG, G.R. PETTIT, C.L. HERALD, R. SHORES, and S.H. YUSPA: Bryostatin 1, an activator of protein kinase C, inhibits tumor promotion by phorbol esters in SENCAR mouse skin. Carcinogenesis **8**, 1343 (1987).
50. YUSPA, S.H., H. HENNINGS, T. SAKO, G.R. PETTIT, J. HARTLEY, and P.M. BLUMBERG: Tumor Promotion: A Problem of Differential Responses of Normal and Neoplastic Cells to Trophic Stimuli. In: Anticarcinogenesis and Radiation Protection (CERUTTI, P.A., O.F. NYGAARD, and M.G. SIMIC, eds.), p. 169. Plenum Pub. Corp. 1987.
51. GSCHWENDT, M., G. FURSTENBERGER, S. ROSE-JOHN, M. ROGERS, W. KITTSTEIN, G.R. PETTIT, C.L. HERALD, and F. MARKS: Bryostatin 1, an activator of protein kinase C, mimics as well as inhibits biological effects of the phorbol ester TPA *in vivo* and *in vitro*. Carcinogenesis **9**, 555 (1988).
52. DELL'AQUILA, M.L., H.T. NGUYEN, C.L. HERALD, G.R. PETTIT, and P.M. BLUMBERG: Inhibition by Bryostatin 1 of the Phorbol Ester-induced Blockage of Differentiation in Hexamethylene Bisacetamide-treated Friend Erythroleukemia Cells. Cancer Res. **47**, 6006 (1987).
53. KISS, Z., E. DELI, P.R. GIRARD, G.R. PETTIT, and J.F. KUO: Comparative Effects of Polymyxin B, Phorbol Ester and Bryostatin on Protein Phosphorylation, Protein Kinase C Translocation, Phospholipid Metabolism and Differentiation of HL60 Cells. Biochem. Biophys. Res. Comm. **146**, 208 (1987).
54. GARRISON, J.C., G.R. PETTIT, and E.M. UYEKI: Effect of Phorbol and Bryostatin I on Chondrogenic Expression of Chick Limb Bud, *in vitro*. Life Sciences **41**, 2055 (1987).
55. PASTI, G., E. RIVEDAL, S.H. YUSPA, C.L. HERALD, G.R. PETTIT, and P.M. BLUMBERG: Contrasting Duration of Inhibition of Cell–Cell Communication in Primary Mouse Epidermal Cells by Phorbol 12,13-Dibutyrate and by Bryostatin 1. Cancer Res. **48**, 447 (1988).

56. DELL'AQUILA, M.L., C.L. HERALD, Y. KAMANO, G.R. PETTIT, and P.M. BLUMBERG: Differential Effects of Bryostatins and Phorbol Esters on Arachidonic Acid Metabolite Release and Epidermal Growth Factor Binding in C3H 10T1/2 Cells. Cancer Res. **48**, 3702 (1988).

57. WARREN, B.S., Y. KAMANO, G.R. PETTIT, and P.M. BLUMBERG: Mimicry of Bryostatin 1 Induced Phosphorylation Patterns in HL-60 Cells by High Phorbol Ester Concentrations. Cancer Res. **48**, 5984 (1988).

58. DE VRIES, D.J., C.L. HERALD, G.R. PETTIT, and P.M. BLUMBERG: Demonstration of Sub-Nanomolar Affinity of Bryostatin 1 for the Phorbol Ester Receptor in Rat Brain. Biochem. Pharmacol. **37**, 4069 (1988).

59. PARKER, J., M. WAITE, G.R. PETTIT, and L.W. DANIEL: Stimulation of arachidonic acid release and prostaglandin synthesis by bryostatin 1. Carcinogenesis **9**, 1471 (1988).

60. JETTEN, A.M., M.A. GEORGE, G.R. PETTIT, C.L. HERALD, and J.I. REARICK: Action of Phorbol Esters, Bryostatins, and Retinoic Acid on Cholesterol Sulfate Synthesis: Relation to the Multistep Process of Differentiation in Human Epidermal Keratinocytes. J. Invest. Dermatol. **93**, 108 (1989).

61. WENDER, P.A., C.M. CRIBBS, K.F. KOEHLER, N.A. SHARKEY, C.L. HERALD, Y. KAMANO, G.R. PETTIT, and P.M. BLUMBERG: Modeling of the Bryostatins to the Phorbol Ester Pharmacophore on Protein Kinase C. Proc. Nat. Acad. Sci. (USA) **85**, 7197 (1988).

62. TRENN, G., G.R. PETTIT, H. TAKAYAMA, J. HU-LI, and M.V. SITKOVSKY: Immunomodulating Properties of a Novel Series of Protein Kinase C Activators. J. Immunology **140**, 433 (1988).

63. MOHR, H., G.R. PETTIT, and A. PLESSING-MENZE: Co-Induction of Lymphokine Synthesis by the Antineoplastic Bryostatins. Immunobiol. **175**, 420 (1987).

64. HESS, A.D., M.K. SILANSKIS, A.H. ESA, G.R. PETTIT, and W.S. MAY: Activation of Human T Lymphocytes by Bryostatin. J. Immunology **141**, 3263 (1988).

65. GOLDE, D.W., and J.C. GASSON: Hormones that Stimulate the Growth of Blood Cells. Scient. Amer. **1988**, July, p. 62.

66. LAVER, J., and M.A.S. MOORE: Clinical Use of Recombinant Human Hematopoietic Growth Factors. J. Nat. Cancer Inst. **81**, 1370 (1989).

67. METCALF, D.: Haemopoietic growth factors: Therapeutics. Med. J. Australia **148**, 516 (1988).

68. MAY, W.S., S.J. SHARKIS, A.H. ESA, V. GEBBIA, A.S. KRAFT, G.R. PETTIT, and L.L. SENSENBRENNER: Antineoplastic bryostatins are multipotential stimulators of human hematopoietic progenitor cells. Proc. Nat. Acad. Sci. (USA) **84**, 8483 (1987).

69. LEONARD, J.P., W.S. MAY, J.N. IHLE, G.R. PETTIT, and S.J. SHARKIS: Regulation of Hematopoiesis IV: The Role of Interleukin-3 and Bryostatin 1 in the Growth of Erythropoietic Progenitors from Normal and Anemic W/W Mice. Blood **72**, 1492 (1988).

70. FIELDS, A.P., G.R. PETTIT, and W.S. MAY: Phosphorylation of Lamin B at the Nuclear Membrane by Activated Protein Kinase C. J. Biol. Chem. **263**, 8253 (1988).

71. MCBAIN, J.A., G.R. PETTIT, and G. MUELLER: Bryostatin 1 Antagonizes the Terminal Differentiating Action of 12-O-Tetradecanoylphorbol-13-acetate in a Human Colon Cancer Cell. Carcinogenesis **9**, 123 (1988).

72. KISS, Z., E. DELI, M. SHOJI, H.P. KOEFFLER, G.R. PETTIT, W.R. VOGLER, and J.F. KUO: Differential Effects of Various Protein Kinase C Activators on Protein Phosphorylation in Human Acute Myeloblastic Leukemia Cell Line KG-1 and Its Phorbol Ester-resistant Subline KG-1a. Cancer Res. **47**, 1302 (1987).

73. STONE, R.M., E. SARIBAN, G.R. PETTIT, and D.W. KUFE: Bryostatin 1 Activates Protein Kinase C and Induces Monocytic Differentiation of HL-60 Cells. Blood **72**, 208 (1988).
74. KRAFT, A.S., F. WILLIAM, G.R. PETTIT, and M.B. LILLY: Varied Differentiation Responses of Human Leukemias to Bryostatin 1. Cancer Res. **49**, 1287 (1989).
75. JETTEN, A.M., M.A. GEORGE, G.R. PETTIT, and J.I. REARICK: Effects of Bryostatins and Retinoic Acid on Phorbol Ester- and Diacylglycerol-induced Squamous Differentiation in Human Tracheobronchial Epithelial Cells. Cancer Res. **49**, 3990 (1989).
76. DALE, I.L., T.C. BRADSHAW, A. GESCHER, and G.R. PETTIT: Comparison of Effects of Bryostatins 1 and 2 and 12-O-Tetradecanoylphorbol-13-acetate on Protein Kinase C Activity in A549 Human Lung Carcinoma Cells. Cancer Res. **49**, 3242 (1989).
77. DREXLER, H.G., S.M. GIGNAC, R.A. JONES, C.S. SCOTT, G.R. PETTIT, and A.V. HOFFBRAND: Bryostatin 1 Induces Differentiation of B-Chronic Lymphocytic Leukemia Cells. Blood **74**, 1747 (1989).
78. DREXLER, H.G., S.M. GIGNAC, G.R. PETTIT, and A.V. HOFFBRAND: Synergistic Action of Calcium Ionophore A23187 and Protein Kinase C activator Bryostatin 1 on Human B Cell Activation and Proliferation. Eur. J. Immunology **20**, 119 (1990).
79. ADAMSON, R.H., B. CHABNER, and H. FUJIKI: U.S.–Japan Seminar on "Marine Natural Products and Cancer". Jpn. J. Cancer Res. **80**, 1141 (1989).

(*Received June 1, 1990*)

Author Index

Page number printed in *italics* refer to References

Subject Index

Fortschritte der Chemie organischer Naturstoffe

Progress in the Chemistry of Organic Natural Products

Volume 56:

1991. 8 figures. X, 188 pages. Cloth DM 220,–, öS 1540,–.
ISBN 3-211-82188-0

Contents: J. Asselineau: Bacterial Lipids Containing Amino Acids or Peptides Linked by Amine Bonds. – J. Kagan: Naturally Occurring Di- and Trithiophenes.

Volume 55:

1989. 41 figures. X, 208 pages. Cloth DM 190,–, öS 1330,–.
ISBN 3-211-82087-6

Contents: M. T. Davies-Coleman and D. E. A. Rivett: Naturally Occurring 6-substituted 5,6-dihydro-α-pyrones – K. Krohn: Building Blocks for the Total Synthesis of Anthracyclinones – M. Lounasmaa and J. Galambos: Indole Alkaloid Production in Catharanthus Roseus Cell Suspension Cultures – C. E. James, L. Hough, and R. Khan: Sucrose and Its Derivatives.

Volume 54:

1988. VII, 353 pages. Cloth DM 320,–, öS 2240,–.
ISBN 3-211-82086-8

Contents: T. Murakami and N. Tanaka: Occurrence, Structure and Taxonomic Implications of Fern Constituents.

Volume 53:

1988. 72 figures. VIII, 311 pages. Cloth DM 275,–, öS 1930,–.
ISBN 3-211-82074-4

Contents: L. F. Alves: Chemical Ecology and the Social Behavior of Animals – T. Nomura: Phenolic Compounds of the Mulberry Tree and Related Plants – A. Chimiak and M. J. Milewska: N-Hydroxyamino Acids and Their Derivatives.

Volume 52:

1987. 65 figures. VIII, 224 pages. Cloth DM 210,-, öS 1470,-.
ISBN 3-211-81989-4

Contents: U. Weiss, L. Merlini, and G. Nasini: Naturally Occurring Perylene-quinones - H. Achenbach: The Pigments of the Flexirubin-Type. A Novel Class of Natural Products - T. Goto: Structure, Stability and Color Variation of Natural Anthocyanins - P. Bhattacharyya and D. P. Chakraborty: Carbazole Alkaloids.

Volume 51:

1987. VII, 317 pages. Cloth DM 280,-, öS 1960,-.
ISBN 3-211-81972-X

Contents: M. Gill and W. Steglich: Pigments of Fungi (Macromycetes).

Volume 50:

1986. 71 figures. IX, 261 pages. Cloth DM 210,-, öS 1470,-.
ISBN 3-211-81969-X

Contents: L. Jaenicke and F.-J. Marner: The Irones and Their Precursors - M. Lounasmaa and P. Somersalo: The Condylocarpine Group of Indole Alkaloids - U. Séquin: The Antibiotics of the Pluramycin Group (4*H*-Anthra [1,2-*b*]pyran Antibiotics) - R. M. Wenger: Cyclosporine and Analogues - Isolation and Synthesis - Mechanism of Action and Structural Requirements for Pharmacological Activity - H. Inouye and S. Uesato: Biosynthesis of Iridoids and Secoiridoids.

All Volumes and Cumulative Index 1–20 available

Price reduction for subscribers: 10%

Special reduced price (20% reduction) for the complete Series Vols. 1–57 incl. the Cumulative Index to Vols. 1–20

Springer-Verlag **Wien New York**

Sachsenplatz 4-6, A-1201 Wien
175 Fifth Avenue, New York, NY 10010, U.S.A.
Heidelberger Platz 3, D-1000 Berlin 33
37-3, Hongo 3-chome, Bunkyo-ku, Tokyo 113, Japan